NEWTON'S TYRANNY

NEWTON'S TYRANNY

THE SUPPRESSED
SCIENTIFIC DISCOVERIES
OF STEPHEN GRAY
AND JOHN FLAMSTEED

David H. Clark and Stephen P. H. Clark

■■ W. H. FREEMAN AND COMPANY · NEW YORK

Text design by Diana Blume

Cataloging-in-Publication Data available from
the Library of Congress.

Printed in the United States of America

First printing 2000

W. H. FREEMAN AND COMPANY
41 Madison Avenue, New York, NY 10010
Houndmills, Basingstoke RG21 6XS, England

Contents

Preface:
Strange Connections

This book explores the unlikely friendship between the Reverend John Flamsteed, the first Astronomer Royal, and Stephen Gray, a humble dyer and amateur scientist. Their friendship lasted over a quarter century, from the early 1690s until John Flamsteed's death in 1719. The two men were united not only in their love of science but also in a hostile and protracted conflict with Sir Isaac Newton. This book reveals how despite the obstacles Newton placed in their path, Flamsteed and Gray were able to make major contributions to observational astronomy, navigation, electricity, and communications. Their attainment of scientific greatness was achieved despite Newton's tyranny.

Newton was a scientific colossus—the most influential scientist of the age and one of the great figures in history.

His writings represented not just unique contributions to the history of science but also major landmarks in the formation of the ideas and values that have molded our world. On September 12, 1999, *The Sunday Times* (London) named Sir Isaac Newton as its "Man of the Millennium." *Time* magazine, in its analysis of the second millennium, identified Newton as its man of the seventeenth century, with the commendation:

> Newton is the man of the century for this reason: by imagining—and proving—a rational universe, he in effect redesigned the human mind. Newton gave it not only intellectual tools undreamed of before, but with them, unprecedented self-confidence and ambition.

Whether or not Newton deserves these epithets, his intellectual rigor and courage in challenging the boundaries of established knowledge made him an inspirational figure. Newton has come to personify the triumph of scientific reason in the past three hundred years and the victory of knowledge and intellect over ignorance and obscurantism. But behind his genius, Sir Isaac Newton was a deeply complex person who attempted to rule the British scientific community of the early eighteenth century through fear and favor. Flamsteed and Gray were both the victims of this flawed genius, albeit for somewhat different reasons. The story told here will prove to be a more generous judge of the major contributions to science of Flamsteed and Gray than was Newton.

The interactions between Newton, Flamsteed, and Gray have been re-created largely from their exchange of letters and through a few written anecdotes of others. Copious quotations from the letters are made in an attempt to capture the authentic nature of their interactions. Although Newton had contemporary biographers, the other two did not. Newton's hagiographers overlooked the extremes of behavior that characterized his interactions with those who ever expressed points of view at variance with his own. The contemporary biographies of Newton do not provide a balanced account of the events told here. In eulogizing Newton without attempting to make a balanced assessment of his life, they diminished the role of history. His genius would survive any detailed scrutiny, but the failure to recognize his aggressive character and his tyrannical behavior meant that the genius of others, including Flamsteed and Gray, was not recognized.

Re-creating events when only one side of much of the correspondence has been preserved is not straightforward. Although a good collection of Stephen Gray's letters to Flamsteed has survived, Flamsteed's letters to Gray have not. It is a bit like listening to just one side of a telephone conversation; only part of the story is being revealed to the listener. It is necessary to use knowledge of contemporary events, the established behavior of the individuals involved, and an understanding of circumstances to try to re-create the full story. It needs to be acknowledged that a degree of imagination is required to fill in gaps in understanding. However, this imagination is tempered by the historical realities. This has certainly been the situation in re-creating the background to the friendship between John Flamsteed and Stephen Gray and their harsh treatment at the hands of Isaac Newton.

The story that follows took place in turbulent times. Newton's year of birth, 1642, was a date of infamy in English history. Eleven months before Isaac Newton was born, King Charles I had magnified the friction already existing between the monarch and discordant factions in Parliament by attempting to arrest five members within the House of Commons chamber itself. As Londoners began to rise in an angry mob in response to this bungled coup against Parliament, the king fled London to rally supporters. He sent his wife, Henrietta-Maria, back to her native France to seek aid for the conflict. The long and painful English civil war had begun. King Charles I was beheaded by order of Parliament in 1649, the last case of regicide in Britain, and the Commonwealth government of Oliver Cromwell then ruled until his death in 1658. Cromwell's son Richard briefly succeeded him as Lord Protector. Newton's childhood was shaped by the puritanical era that followed the civil war. Charles II, the son of the murdered monarch, returned from exile in Europe in 1660 to a euphoric welcome from the populace. He was restored to the throne and ruled until his death in 1685.

The rule of King James II, Charles's brother, was brief and characterized by religious turbulence and violent rebellion precipitated by an uprising led by the Duke of Monmouth (an illegitimate son of Charles II). James II was driven into exile in 1688 for trying to force the nation to return to Catholicism, to be replaced by the Protestant William of Orange and his English wife, Queen Mary (James's daughter, who had retained her Protestant faith). Mary predeceased her husband. On William's death in 1702, following a riding accident, Queen Anne (Mary's sister, who had also remained Protestant) came to the throne, to be succeeded on her death in 1714 by King George I of Hanover (a very distant relative of the former Stuart monarchs). When King George I died in 1727, the same year as Newton's death, King George II succeeded him. Each of these monarchs, from Charles II to George II, played a role in the evolving saga of Newton, Flamsteed, and Gray. Royal patronage—and royal control—changed the course of this history of science.

Isaac Newton was quick to use royal patronage to his advantage. However, he was unwilling to recognize the legitimate intellectual contributions of others. Newton's well-documented conflicts with the German philosopher-mathematician Gottfried Wilhelm von Leibniz over the origin of the calculus and with the English physicist Robert Hooke over the nature of gravity and of light were typical of his hostile responses to the intellectual achievements of others. The origins of his insecurity can be traced to a troubled childhood, but enormous success in his adult life might have been expected to produce a more generous spirit. Newton, however became increasingly mean-spirited as his scientific stature grew. Nowhere is this more apparent than in his dealings with Flamsteed. In his monumental treatise on the nature of gravitation, the *Principia*, Newton drew heavily on the careful astronomical observations of Flamsteed; yet in later editions of the work Newton systematically expunged the references to Flamsteed's contributions. Even by the standards of the well-documented battles between Newton and other eminent scientists of the day, the hatred between Flamsteed and Newton was particularly intense.

The story of Flamsteed and Gray and their relationships with Newton is full of intrigue, of uncontrolled ambition, and of the conflict of egos. Flamsteed had the confidence and scientific stature to challenge the authority of Newton. Gray did not. But the mutual respect between Flamsteed and Gray meant that Newton slighted Gray for no greater reason than his friendship with the Astronomer Royal. Noble endeavor was countered by claims of unscrupulous plagiarism; personal friendships were strained through rumor and innuendo.

The story of Gray and Flamsteed is one of loyalty and commitment against great odds. But it is also the story of the significant scientific advances made by the two men. Out of personal turmoil, truth eventually triumphed.

In 1729, two years after the death of Newton, Stephen

Gray carried out a startling experiment in electrical communication. He demonstrated that electrical phenomena could be transmitted in a controlled fashion over a distance of hundreds of yards. He was the first person to witness this unexpected communication effect. If Alexander Graham Bell, the inventor of the telephone in 1876, can be called the "father of communication," then Stephen Gray deserves to be acclaimed as the "great-grandfather of communication," since it was Gray's experiment of 1729 that first demonstrated electrical communication over a considerable distance. It is this epoch-defining electrical transmission experiment of 1729 that represents the climax of our story. It defined Gray's ultimate intellectual triumph over the tyranny of Newton and the decades of scientific neglect. Flamsteed's classic astronomical work had been published posthumously just four years earlier, ensuring his scientific immortality despite the indignities he suffered during his lifetime at the hands of Newton.

Lifestyles today are being transformed by the modern communications revolution. The inhabitants of the global village keep in touch with mobile telephones and fax machines. The Internet is revolutionizing the way information is exchanged, and e-mail is for many now the preferred means of communication. Multimedia technology is redefining education and entertainment. We are today the ultimate beneficiaries of the pioneering research of Faraday, Babbage, Morse, Bell, Edison, Marconi, and many other well-known scientists from the past three centuries.

The genealogy, however, is clear: the communications revolution actually had its humble origins in the tentative scientific experiments of a forgotten man of science, Stephen Gray. This unsung hero was the first to experience the early birth pangs of the communications revolution.

Satellite navigation systems now allow us to locate a position anywhere on the Earth's surface to within a few meters. Post-Renaissance astronomy has transformed our understanding of the cosmos, and we now recognize that Earth is a very typical planet orbiting an ordinary star, one of billions of stars within a very conventional galaxy (the Milky Way). The Milky Way is one of billions of galaxies making up the visible universe. The universe is undergoing violent upheaval, and planet Earth is like a mere speck of sand swept by the tides of universal change. Our new understanding of the cosmos and the new systems we now use for precision navigation are constructed on the secure foundations put down by John Flamsteed.

And as for Isaac Newton: well, he redefined the path of science in a manner that no one had done before and only Albert Einstein has done since. The new understanding he brought to the nature of matter, forces, light, and gravity set science in a new direction that took it within 150 years to the industrial revolution—and thence to the technological world in which we now live. But greatness cannot disguise Newton's tyranny.

Acknowledgments

We wish to thank our agent, Al Zuckerman of Writers House, and our editor, John Michel of W. H. Freeman and Company, for their excellent advice and assistance during the writing of this book.

One of us (DHC) was introduced to the correspondence of Stephen Gray by Phillip Laurie, who was the archivist at the Royal Greenwich Observatory until 1976. His successor, Janet Dudley, offered valuable guidance. Much of the early background research on Gray's manuscripts was carried out in collaboration with Lesley Murdin in 1978 and 1979. Valuable assistance in tracking down archive material was provided by staff at the British Museum, Canterbury Cathedral, the Royal Society, the University Library, Cambridge, and the Bodleian Library, Oxford. Thanks must be given to the historians of Christ Church, Oxford, and the Faculty of Education, Cambridge.

Much of this book is based on manuscript letters between the principal characters, and excerpts from these letters are used freely. Dr. Samuel Johnson's great English dictionary had not yet been produced at the time covered by the letters, and the spelling and punctuation they contain is sometimes idiosyncratic (especially in the case of Stephen Gray). We have used modern spelling in most cases and inserted some additional punctuation to aid understanding, while taking great care to preserve the writer's intended meaning.

We are especially indebted to our family for all their encouragement and significant help during the long hours of writing. Suzanne Clark researched information on the historic sites.

David Clark and Stephen Clark
October 2000

CHAPTER 1

The Complex Paths to Truth

I do not know what I may appear to the world, but to myself I seem to have been only like a boy playing on the seashore, and diverting myself in now and then finding a smoother pebble or a prettier shell than ordinary, whilst the great ocean of truth lay all undiscovered before me.

—ISAAC NEWTON

It is a short but picturesque boat journey from the city of London down the river Thames to the village of Greenwich. However, it was a somewhat unusual and unexpected journey for Sir Isaac Newton to be making on a bitterly cold spring morning in April 1704. He was the man who had revealed the nature of gravity, invented the calculus, unlocked the secrets of light, and brought a new understanding to the forces that control the universe. As one of the most famous individuals in England, Newton could have sent an entire retinue to Greenwich in his stead. But this was as much a personal as a professional matter, and he felt compelled to make the journey himself.

Newton was acclaimed by his supporters to be the finest philosopher of this or any previous age. John Locke, a genius himself, referred to "the incomparable Mr. Newton." The simple myth that truth had been revealed to him by a falling apple added to an aura of veneration surrounding the great man. Such was his fame that the king of England had appointed him to one of the most important and highly paid positions in the land. Master of the Royal Mint, Newton now controlled England's supply of money. This position of eminence and authority was a far cry from his humble beginnings as an academic at Trinity College, Cambridge University. There he had worked in self-imposed isolation to understand the mysteries of the natural world. As a lonely academic, Newton had detached himself from the mundane affairs of life. He took no care of himself. He ate little and irregularly, finding food an unwelcome distraction from his intellectual endeavors. He seldom went to bed until the small hours of the morning, and he took no recreation. He went about, as a contemporary noted, "with shoes down at heels, stockings untied, and his head scarcely combed." Even romantic liaisons held no interest for him.

The French philosopher Voltaire commented famously on Newton's sexuality:

> Sir Isaac, during the long course of years he enjoyed, was never sensible to any passion, was not subject to the common frailties of mankind, nor ever had any commerce with women.

Vengeance

As an academic Newton was the archetypal "absent-minded professor." The position as Master of the Royal Mint was to change his lifestyle in almost every respect. Social interactions, food, and dress, for so long unwelcome distractions, were now all of importance to the "new" Newton. A knighthood from Queen Anne followed, and he had taken on his new mantle of eminence with style, commitment, and dignity. His journey to Greenwich this spring morning, however, had nothing to do with his official responsibilities at the mint or his prominent new position in London society. Its purpose was entirely sinister. The great Sir Isaac Newton was traveling to Greenwich to seek revenge, determined to rekindle the embers of an old feud.

Inspiration is rarely stimulated by vengeance; it is a destructive force, not a creative one. Despite his great intellect, Isaac Newton had never been able to grasp this basic fact. He had suffered a troubled childhood, the likely origin of his tyrannical traits. His natural father, also named Isaac, was a yeoman farmer who had died just two months before the baby's premature birth. His widowed mother, Hannah, had remarried a clergyman, Barnabas Smith, when Isaac was just three years old. But rather than taking him with her to the new marital home, Hannah had left her son in the care of her parents. Newton never forgave this act of desertion and maternal betrayal, and it was probably the cause of the insecurity and mistrust of others that plagued his whole life. The young Newton hated his stepfather, and he had made references to this hatred in his youthful writings. When at age twenty Newton had recorded in a notebook a list of sins in his life so far for which he should do penance, he included, "threatening my father and mother Smith to burn them and the house over them." The hurt of his early neglected childhood had lingered with him, and in his adult life he found it difficult to form close relationships or to display affection. His temper and intolerance, characteristics of childhood, had become legion in his later life when a cruel temperament was used to devastating effect. Revenge was to become a habit in his professional life. In seeking retribution Newton would actually inhibit scientific progress by others, even as he made monumental progress himself.

At sixty-one, Newton was too old to ride a horse, and though he could have ordered a carriage to convey him to Greenwich, the route was still a partly muddy, partly frozen cart-laden track that could never be traveled in comfort. It had been a long harsh winter, and the spring was late in coming. Europe had been trapped for decades in a mini–Ice Age, and for much of the winter the Thames had been frozen over, although it had thawed now. Hence he had opted for the relatively tranquil river passage.

Tall ships and schooners were berthed alongside warehouses, landing goods from ports in Europe, the Indies, and the New World. Smaller river and coastal craft and larger naval vessels confirmed that here was the heart of the nation's trade. Newton's London was one of the world's great trading capitals. As trade soared England's merchants prospered, and so too did the Master of the Royal Mint: Newton's income depended on the coinage pressed, and a dynamic economy required more coinage. Thus, in addition to his salary of £600 a year, he received payment for every pound of silver or gold coined—on average an extra income of £1,000 each year.

On reaching the landing at the Greenwich royal park, Newton was forced to follow the frozen track across the park and up the hill overlooking the river. His long wig, tricorne hat, leather breeches, and ankle-length coat provided some protection against the penetrating cold, but he was somewhat portly now and could maintain only a sedate pace. Pausing halfway to the top, he looked down briefly over the splendid Palladian mansion at the river's edge, once the home of Henrietta-Maria, widow of King Charles I.

Across and up river from the Queen's House, Newton could see the slums of East London, home to the coin clippers and counterfeiters. One of his duties at the mint had been to track them down and send them to the gallows. Looking farther up river, he could admire the grandeur of London. Although two-thirds of the city had been destroyed in the catastrophic fire of 1666, it had since been rebuilt. And if it had cost lives and property, at least the fire had rid the city of the rats that had brought the great plague. The plague had decimated the population the year before the fire. Infected houses were bricked up, with whole families left trapped inside to die.

Ironically, the plague year had been a golden period of creativity for Newton. Although the epidemic had been confined largely to London, and in particular to those crowded and unsanitary parts of the city where the poorest lived, Cambridge University had also been closed following the outbreak. It was a sensible precaution because of the number of students coming from the capital. Many of them had retired to the countryside to continue studying with their tutors. Newton, as always, preferred to be alone. He had returned to his mother's home to contemplate all that he had learned during his university studies and to carry out various experiments in his quest to come to terms with the natural world. He had conducted his first experiments on the nature of light in his study at his mother's home, Woolsthorpe Manor. And it was during this time, in the garden at Woolsthorpe, that he had sat under an apple tree contemplating the nature of gravitation. He later commented on his creativity at this time:

> All this was in the two years of 1665 and 1666, for in those years I was in the prime of my age for invention and minded Mathematics and Philosophy more than at any time since.

But now was not the time for further contemplation. He took one final glimpse back toward the city. In the distance Saint Paul's Cathedral now dominated the London skyline. The architectural masterpiece of his good friend Sir Christopher Wren was still engulfed in scaffolding as it neared completion, but the majesty of its grand dome was already apparent. Newton turned and headed toward another elegant Wren building, the Royal Observatory, which had been built on the orders of King Charles II in 1675.

Newton was visiting an old adversary—the Reverend John Flamsteed, who had been appointed the first Astronomer Royal by Charles II. Although Newton and Flamsteed had known each other for thirty years and had once been close, the two men had barely communicated for the past eight years. In the years from 1694 to 1696, Flamsteed and Newton had collaborated in trying to understand the complex orbit of the Moon around the Earth. Newton saw a theory of the Moon's motion as an ultimate validation of his famous law of gravitation and a means of solving a major problem of the times—navigation at sea. It was believed that the answer to the problem lay in the heavens. And it was in the heavens that Newton had achieved his greatest scientific success. Observations of the planets and comets, which had helped Newton demonstrate the value of his law of gravitation, had come from Flamsteed. Newton had depended on them for his monumental treatise *Philosophiae Naturalis Principia Mathematica,* one of the most influential books ever published. In the years while they were still on amicable terms, Flamsteed had also provided Newton with an ample supply of accurate observations of the Moon's path across the sky. But despite the data from Flamsteed, a satisfactory theory of the Moon's motion had still eluded Newton. He had demanded more and more observations from Flamsteed and had grown frustrated when all that he required were not forthcoming, at least in the form and with the accuracy he desired. Newton finally gave up the challenge in 1696, when he moved from his academic appointment in Cambridge to take up the new responsibilities at the mint in London. But his failure to produce a theory of the Moon's motion still haunted him. He had convinced himself that Flamsteed was to blame for his earlier failure, and he had been unable to accept that the problem might lie, at least in part, in the limitations of his own understanding. Newton now looked on Flamsteed as an enemy. It was therefore a strange decision by Newton to renew contact with him now.

Power, Plot, and a Pretty Niece

Just five months before Newton's visit to Greenwich, he had been elected president of the Royal Society. The society was then, as it has remained since, the bastion of science in Britain. The Royal Society had had its origins in the infrequent meetings of "the philosophically minded" in the rooms of John Wilkins, a fellow of Wadham College, Oxford. The meetings of what became known as the "Invisible College" commenced in about 1648. In 1659 the "Invisible College" acquired a regular meeting room in Gresham College, in Bishopsgate, London. (A wealthy merchant, Sir Thomas Gresham, founded Gresham College in 1575 to offer education to the citizens of London.) Meetings of the "Invisible College" became formalized, centered on the reading of papers. The king, Charles II, had a fascination with all things philosophical. He was an occasional attendee at meetings at Gresham College, and in 1662 he granted the "Invisible College" a royal charter; it became the "Royal Society." The royal motto given to the society was *Nullius in Verba*—"Take no one's word for it."

The presidency of the Royal Society was a position of considerable influence and power, and Newton intended to use this power to favor those who had supported him and to suppress those who had not. He now placed Flamsteed in the latter category.

John Flamsteed viewed the pending arrival of his illustrious guest this April morning with deep foreboding. He had held the appointment of Astronomer Royal for almost thirty years and, despite persistent ill health, had dedicated that time to conducting careful and accurate observations of the heavens. Flamsteed was widely admired in the world of astronomy, and he had been saddened to earn the wrath of the great Sir Isaac Newton. Having been informed by penny post a few days earlier that Newton planned to visit about one hour before noon, Flamsteed was now waiting in the upper Octagon Room of the observatory looking out over Greenwich Park. The splendid Octagon Room had been designed by Wren to house the telescopes and clocks needed for precision observations of the heavens. On a normal day, Flamsteed would have been asleep at this time following a full night's observing. His disposition was not improved by being deprived of much-deserved sleep. Nevertheless, catching his first glimpse of Newton climbing the hill, he went out to greet him.

Flamsteed was four years Newton's junior, but in physical appearance he looked very much older. He had suffered poor health since childhood and was now crippled with gout. He had been born in 1646 in the village of Denby, near Derby in the English Midlands. Like Newton, Flamsteed also had suffered a troubled childhood. His mother had died when he was just three years old, and he never forgave his father for hurrying into remarriage. The young Flamsteed studied at the Derby Free School and was noted as a gifted scholar. But when he was fifteen years old, a serious bout of rheumatic fever left him physically weak for life. This burden of physical weakness and persistent ill health led to frustration and a sharp temper. He was humorless and cantankerous, but these unattractive characteristics are hardly surprising in light of the almost constant physical pain he had to endure. Family and friends quickly learned that one should stay well clear of young John when he was suffering from one of his frequent migraine headaches. His father had even sent him to Ireland in search of a miracle cure for his rheumatic limbs. Later, undeterred by failure, Flamsteed visited the faith healer a second time when the man came to Liverpool—with a similar outcome. But being confined to home gave him time to study astronomy, and that rapidly became an obsession. Astronomy provided Flamsteed with an intellectual escape from the misery of persistent illness.

Flamsteed met Newton at the observatory gate. The greeting between the two old adversaries lacked warmth, but they each observed the standard courtesies, inquiring after each other's health and commenting on the severity of the frost. Flamsteed escorted his guest to the domestic quarters beneath the observatory, where Mrs. Flamsteed was preparing coffee. Flamsteed had married late in life, when he was forty-eight. Née Margaret Cooke, the daughter of a lawyer, his wife was eight years his junior and the granddaughter of a previous rector of the village of Burstow where Flamsteed himself was now the rector. Flamsteed's remuneration as Astronomer Royal was a modest one, and to earn sufficient money to buy instruments and employ assistants he had had to take the position of rector at Burstow in 1684. Since Burstow was some thirty miles south of Greenwich, he could visit the parish only for a few days every month or two, and he employed a curate who could conduct the services on most Sundays. Harvest time in particular was a time when Flamsteed always spent several weeks at Burstow, not only seeing to the pastoral needs of his parishioners but also overseeing the collection of their tithes. As well as being a gifted astronomer and dedicated cleric, John Flamsteed was a canny businessman. The Burstow tithes amounted to £153 a year, from which the curate was paid £40.

After serving coffee and adding more wood to the fire, Mrs. Flamsteed excused herself. She was an intelligent woman with scientific interests herself, and she served her husband as a diligent assistant in his astronomical observations, frequently sitting through cold nights with him recording his measurements. Mrs. Flamsteed was fiercely protective of her husband's work and shared his concern about the possible purpose of Newton's visit.

The purpose of Newton's visit was to insist that Flamsteed should publish the results of his thirty years of observing at Greenwich. Flamsteed was a perfectionist, however, and did not wish to publish his results until the heavens had been completely surveyed and he could produce a definitive catalogue of the stars. But Newton needed Flamsteed's data now in order to produce a theory of the Moon's motion that he could include as the centerpiece of a new edition of the *Principia*. Newton was convinced that Flamsteed possessed observations that would enable him to develop a valid theory for the Moon's motion and that Flamsteed had deliberately withheld these observations in their earlier interactions. Newton believed that Flamsteed already held the key to the final validation of his grand theory of gravity and to the solution of the problem of navigation at sea. The reality was that Flamsteed had already passed to Newton all the lunar data he genuinely felt were valid; there was no withheld treasure chest of lunar data. But so deep were Newton's suspicions that he would never accept this fact.

Over the preceding months Newton had developed a Machiavellian plan to extract from Flamsteed all of his data. He knew he could get royal support to force an early publication of Flamsteed's results—or at least those lunar and planetary observations he needed to complete a new version of the *Principia*. As a second part of his plan, Newton was working with others to get a new astronomical observatory established at his old Cambridge college, Trinity, that could challenge the status of Flamsteed at Greenwich as the prime source of astronomical observations.

Newton's confidence in securing royal support for an early publication of Flamsteed's results was based on a rather unusual family connection to the inner sanctums of the court. Newton had an attractive young niece by the name of Catherine Barton, the daughter of one of Newton's stepsisters from his mother's marriage to Barnabas Smith.

When Newton moved to London, Catherine was sent to housekeep for her uncle. Such was her beauty and flirtatious wit that she acquired many suitors among the social visitors to her uncle's house. One of these was Baron Halifax. Halifax was Chancellor of the Exchequer (the government minister in charge of the treasury) and was as close to the monarch as anyone in the kingdom. Newton had known Halifax at Cambridge as Charles Montague, and it was Montague who had arranged the position for him at the mint. After leaving Cambridge, Montague had risen quickly to become a political star, and King William had made him a baron. Halifax was fat, forty, and far from handsome, but he was rich and powerful, and he became infatuated with Newton's niece. By 1703 he and Catherine Barton were lovers. Even Voltaire, who promoted Newton and the importance of his work throughout Europe, was later to observe:

> In my youth I thought that Newton had made his fortune by his great merit. I had supposed that the Court and the City of London had named him Master of the Mint by acclamation. Not at all. Isaac Newton had a very charming niece: she greatly pleased the Lord of the Treasury, Halifax. The infinitesimal calculus and gravity would have availed nothing without a pretty niece.

The Halifax-Barton affair scandalized London society, and with his puritan upbringing Newton could hardly have condoned it. Nevertheless he was prepared to use this unlikely family connection when it suited his purposes.

From Halifax, Newton had established that Prince George of Denmark, the consort to the newly crowned Queen Anne, was interested in astronomy. Halifax further ascertained that Prince George would be willing to support the early publication of Flamsteed's work. Indeed, he would make the monies available to pay for the publication. He would also agree that the Royal Society should be given the job of overseeing the publication. What could be more natural than the work of the Astronomer Royal being published with royal patronage under the control of the Royal Society (with Newton as its president)? The scheme seemed to be perfect. How could Flamsteed, as the Astronomer Royal, possibly refuse such royal interest in his work?

Newton's manner was one of forced courtesy during the discussions with the Astronomer Royal. He inquired how Flamsteed's catalogue of the stars was progressing. The survey had already taken almost thirty years, so it was not unreasonable to assume that it was nearing completion. Newton raised the issue of how quickly Flamsteed might make his results available for printing, should royal support be forthcoming. Flamsteed, not wishing to display any misgivings about his own work (although much still remained to be done to bring his observations to his desired state of perfection), gave a reasonably encouraging reply. And he was flattered to learn that Newton was prepared to seek royal funding for the publication. Newton agreed to write with further proposals on how progress should be made.

Mrs. Flamsteed interrupted the lengthy conversation, returning to inquire whether Newton would be staying for a meal. He replied that he would take some more coffee and eat a little cold meat with some bread. However, he did not have time for a more substantial meal, since he had arranged for a boatman to return to collect him an hour after noon.

And so Newton bade him farewell, urging Flamsteed to "do all the good in your power." Flamsteed assured him that that had always been his intent, but caustically noted in his record of the meeting, "though I do not know that it ever has been of his."

Flamsteed returned to the Octagon Room and peered after the retreating figure walking down the hill toward the river. He could see no good coming from the visit and he was right to be concerned. Newton's plans would precipitate a series of events that would set science back by decades.

Newton, however, reflecting on the events of the morning during the boat journey back to the mint, was well pleased. His plans were taking shape. This time Flamsteed would be forced to do his will. There would be no escape for the irascible Astronomer Royal.

Canterbury Tales

Some sixty miles southeast of Greenwich is the ecclesiastical city of Canterbury, dominated by its magnificent cathedral. Although there has been a monastery on the site since the seventh century, the cathedral dates from Norman times. Walking around this remarkable edifice is like stepping back into the history books and into the monumental events of the Middle Ages. For centuries following the murder of Archbishop Thomas Becket on December 29, 1170 (supposedly at the behest of King Henry II), the cathedral was a site of Christian pilgrimage second in popularity only to Rome. Travelers flocked to the shrine where four years later Henry did penance for his treachery against Becket. The stone steps to the martyr's shrine display the wear of millions of pilgrims approaching on their knees. The city had scores of inns and taverns for the pilgrims, and the atmosphere of pilgrimage was immortalized in Chaucer's *Canterbury Tales*.

With King Henry VIII's break with Rome and his dissolution of the monasteries, the cult of Becket was discouraged and the city of Canterbury fell into decline. In the seventeenth century it reemerged as a center of commerce, in part due to an influx of French Protestants fleeing the persecution of Louis XIV. Among the refugees were skilled silk weavers, and Canterbury soon became a major silk center. Dyers were needed for the new trade.

A dyer of Canterbury would unexpectedly be caught up in the consequences of Newton's visit to Greenwich in 1704. His name was Stephen Gray. Gray was a keen amateur scientist. The great Sir Isaac Newton would not normally associate with someone of so lowly a social station as a dyer, but Gray was a friend and scientific collaborator of John Flamsteed and would soon become enmeshed in Newton's plans to call the Astronomer Royal to account. Newton had no direct argument with Gray. But Gray would stand beside Flamsteed throughout his feud with Newton, and for Newton the friend of an enemy was also his enemy—it was as simple as that.

Mathias Graye, the son of a blacksmith, had married Anne Tilman in July of 1658. He had not only changed the family trade to dyeing, he had also dropped the letter *e* from the family name. Mathias and Anne had seven children, four boys and three girls. The oldest boy, born the year following the marriage, was named Thomas. The second son was named Mathias, after his father. Then followed two daughters, Sarah and Elizabeth. Stephen was the third boy. Then came John and finally Mary.

Canterbury in the mid-seventeenth century was a city of fewer than three thousand inhabitants within the city wall. It was a busy market center, with ecclesiastical attractions. It was a city, despite its modest size, by virtue of having a cathedral. Stephen Gray was born in 1666. This was the year of the Great Fire of London that cleared the slums of the rats and brought an end to the plague (allowing Isaac Newton to return to Cambridge). Stephen Gray was baptized on December 26, 1666 at All Saints Church in Best Lane, over three months after the Great Fire. His date of birth is not recorded, but presumably it was just a few weeks, or perhaps merely days, earlier. Infant mortality was so high during those times that parents rarely lingered in arranging a baptism. The Gray family dyeing business in Best Lane later moved to nearby Stour Street. A river ran behind their shop, providing a ready source of the water needed for dyeing and a handy channel of transport. The seventeenth-century cottages in Best Lane and Stour Street were of the Kentish half-timbered style, a characteristic form of architecture of the period, so solid in the construction of their oak frames as to withstand the weathering of centuries. The narrow streets were all that was needed for horse and carriage. The ritual of disposal of waste into the streets, to be dealt with by the weekly rakers, had supposedly been discontinued by the decree of King Henry IV in 1407, but the streets of seventeenth-century Canterbury were still littered and filthy.

Dyeing and Living

Stephen Gray's youth included many periods of sorrow. His sisters Elizabeth and Mary died before he was ten years old, and his father died when Stephen was just seventeen. The oldest son, Thomas, took over the dyeing business. Their mother, Anne Gray, died just a few years later.

There were two schools in Canterbury at the time. The King's School adjacent to the cathedral was reestablished by King Henry VIII in 1541 on a foundation that dated back probably to the seventh or the eighth century. The Poor Priests Hospital, next door to the Grays' dyeing shop, was an almshouse for the poor with schooling for the children of tradesmen. There is no record of a pupil named Gray in the King's School archives of scholars. Stephen might have attended as a commoner (a fee-paying student, although fees were modest). Education at the King's School centered on the classics (and Stephen Gray did have a sound working knowledge of Latin); there was no mathematics. Gray in later life was comfortable with relatively complicated calculations, suggesting that he was probably educated somewhere other than the King's School (or gained mathematics training elsewhere). The Poor Priests Hospital may be the more likely site of his education. Schooling was far from universal, but Gray senior, establishing himself as a Canterbury artisan, would no doubt have been eager to demonstrate his commitment to the education of his children.

Since Stephen Gray's great-grandfather had purchased the "freedom of the city" to trade as a blacksmith for the princely sum of thirty shillings, his descendants would also enjoy the privilege to trade should they apply for freedom of the city. Stephen followed his brother Thomas into the family dyeing business, and he was awarded the freedom of the city to trade as a dyer in 1692. When Thomas, who was seven years older than Stephen, died in 1695 at age thirty-six, Stephen took control of the dyeing business. The older brother, Mathias, started a grocery business and was later to become mayor of Canterbury (a position of some eminence). The youngest brother, John, was a carpenter. Although his three brothers all married, there is no evidence that Stephen ever took a wife. The only details of his personal circumstances to be gleaned from his letters are his impoverished state and his sickly nature.

Dyeing was a hard life, demanding long hours of toil and contact with a range of harsh chemicals. At this time natural dyes were the only ones used. The dyers, who spent hours with their hands immersed in dye solutions, inevitably suffered from dermatitis. And the toxic fumes given off by the dyes meant that respiratory illnesses were also common. On top of these, lifting the heavy bolts of cloth could cause back strain. So it is not surprising that Gray suffered from ill health throughout his life. In writing to Flamsteed he noted:

> I have been much afflicted with the Dolor Ischiadis for near a quarter of a year but thanks be to God am now almost free from it feeling no pain except I attempt to labour hard.

Several years later he was still complaining of

a strain I received in my back some years ago which brought on me the Dolor Coxendicis. . . . The difficulty and pain caused by my work is more than in former years.

Despite the tough life as a dyer, Stephen Gray found time for science. How Gray developed his love of science is unknown. Possibly it was the chemistry of dyeing that sparked his interest—the mixing of the natural dyestuffs to achieve the desired hue and the treatment of the yarns to enable them to absorb the dyes that generated a fascination with natural materials and phenomena. Silk was to figure in many of Gray's later scientific experiments.

Stephen Gray's early scientific research, pursued late at night after a busy day in the dyeing shop, covered a broad range of interests. Henry Hunt, a boyhood friend from Canterbury, had moved to London to work as an assistant to a scientist named Robert Hooke, who was curator at the Royal Society at the time. (The role of the curator was to prepare experiments and demonstrations for meetings of fellows.) Robert Hooke, one of the most brilliant experimentalists of the seventeenth century, was a man small of stature but large in intellect. Gray used this Canterbury connection to communicate the results of his early experiments to the Royal Society, hoping that his letters to Hunt (and later to other officials of the Royal Society) would be published in the society's *Philosophical Transactions*—the bible of science at the time.

Gray's early letters covered optical experiments, strange properties of certain materials, and some simple astronomical observations. Although there were no results of significance in these early experiments, they do display the attention to detail and great care with experiments that typified Gray's scientific research throughout his life. His experiments were often inspired by the contributions of others in the *Philosophical Transactions*.

Like many of the scientists of the time, Gray became interested in questions of paleobiology. Some bones and teeth were found in a well on the property built for the former husband of the wife of his brother Mathias. The Royal Society sought more information on the find from Stephen Gray. In his reply to the society Gray paints a delightful picture of old Mr. Sommer, the father of the previous owner, being lowered in a basket into the well where the bones were found. Some care was taken in the investigation: a professional "limner," or portrait painter, was hired to make an accurate drawing of the find. It was speculated that the bones might have belonged to an elephant brought over by the Romans, but Gray inclined to Mr. Sommer's explanation that they were the bones of a hippopotamus deposited during the Great Flood!

Gray demonstrated a fascinating ability to utilize his limited experimental resources. For example, in following up reports of observations of "insects" (protozoa) that appeared in long-standing water viewed through a microscope, he extended his observations to all other fluids readily available to him:

I have examined many transparent fluids as water, wine, brandy, vinegar, beer, spittle, urine etc., and do not remember to have found any of these without more or less of the bodies of these insects.

We do not know what generated Stephen Gray's keen interest in astronomy. Gray and Flamsteed seem to have started their correspondence in about 1696, and regular letters between them continued until the later years of Flamsteed's life. Gray's early writings to Flamsteed are businesslike but cordial, possibly indicating that a friendship based on mutual interest in each other's observations had already developed between the two. Flamsteed had a wide circle of people interested in astronomy with whom he corresponded and who provided him with observations as their other interests allowed. Gray, being in trade, differed from most of Flamsteed's collaborators.

Gray's letters record a number of interesting astronomical phenomena. There were observations of both lunar and solar eclipses and spots on the surface of the Sun. There were measurements of the eclipses of Jupiter's satellites, which might prove useful in developing a method for navigation. Gray constructed his own telescopes, including grinding the lenses. He also made microscope lenses for Margaret Flamsteed.

Gray's letters to Flamsteed on astronomical topics often made reference to personal matters. Coming from a younger man, Gray's occasional comments on Flamsteed's health might seem tactless, but they do show concern:

> I am glad to hear you are in some degree of health,
> though as you tell me it is but indifferent for one that
> has been the greatest part of his life time so studious, not
> to say laborious, as you have been and arrived to your age
> cannot expect it much otherwise.

Gray's letters also make frequent reference to his own poor health.

Flamsteed introduced Gray to his wealthy step-nephew, John Godfrey, and Gray carried out some astronomical observations at Godfrey's estate. Family greetings were exchanged in letters to Flamsteed:

> My honoured friend your cousin Godfrey who com-
> mands me to present his humble service to you and
> Madam Flamsteed.

(The term cousin was loosely used for any close relative, and the true relationship is explained by an annotation by Flamsteed on one of Gray's letters referring to Godfrey: "Mr Godfrey and Mr Godfrey that is Member of Parliament for London are own brother's sons." Flamsteed had a step-brother through his father's second marriage; a stepnephew relationship with Godfrey seems the most likely genealogy.)

Gray's normal tone, both in addressing Flamsteed and in writing about him, is one of great admiration. This admiration is so consistent (and repeated in letters to others) that it seems to have been genuine, although admittedly Gray had much to gain from Flamsteed. For many years he hoped to be given a copy of Flamsteed's star catalogue when it should be published, and he referred to existing star catalogues as "vulgar" in comparison with what Flamsteed was producing.

Gray's letters reveal a modest man, lacking confidence and self-respect, who found it difficult to converse with strangers, yet a man who was able to confront great intellectual challenges. He was a gifted experimenter and a careful observer of natural phenomena. His unlikely friendship with John Flamsteed brought together two men separated in age by twenty years, and from very different backgrounds. Gray was able to give Flamsteed the scientific respect he craved; Flamsteed was able to provide the younger Gray with a link to legitimate science that would normally have been beyond someone of so lowly a social station and of such modest means. Only the uncontrolled ambition and flagrant paranoia of Isaac Newton would ultimately stand between the two men and the due recognition of their peers.

By 1704, the year a scheming Isaac Newton visited John Flamsteed at Greenwich, Stephen Gray was thirty-eight years old. He was poor, single, overworked, and in poor health. However, his impoverished and sickly state could not diminish his fascination with natural philosophy. Gray's genius could be suppressed neither by his modest circumstances nor his lack of self-esteem. By this time Gray and Flamsteed were regularly exchanging letters about their scientific ideas, and Flamsteed and Newton were on the verge of a major conflict over Newton's access to the Astronomer Royal's observations. The complex paths toward truth were becoming further intertwined. However, friends and foes had been firmly identified in the events of the previous decades. No good could come from Newton's renewed scheming.

CHAPTER 2

Friends and Foes

Sir Isaac worked with the ore I had dug.
   —JOHN FLAMSTEED

If he dug the ore, I made the gold ring.
   —ISAAC NEWTON

It is not possible to understand fully the consequences of Isaac Newton's visit with John Flamsteed at Greenwich in April 1704 without first understanding certain events that had taken place over the preceding thirty years. At the time, these events seemed to be of limited interest, but taken together they reshaped the course of science. History may appear to be little more than an accident to those immersed in its making. Only when we are able to reflect on earlier events with an understanding of what has come later, can connections be made between seemingly random occurrences.

All at Sea

By the late seventeenth century, navigation at sea had become a major problem. The European nations roamed the oceans, trading, exploring, making war, and pirating one another's vessels. But to make port and to avoid natural perils, good maps and an accurate knowledge of one's position were imperative. As every schoolchild knows, Christopher Columbus in 1492 was so uncertain of his position that he had thought he had reached India rather than having discovered the New World—all because he had no idea how far he had sailed westward—that is, his *longitude*. By the seventeenth century it was generally believed that those who could master navigation at sea could master the world. Generous prizes for accurate navigation systems were offered by the governments of France, Holland, Spain, Portugal, and Venice—and eventually England. Improved astronomical methods were thought to hold the key to accurate navigation.

In the second century B.C., the astronomer Hipparchus had been the first person to define points on a map by an imaginary grid of coordinates (latitude and longitude). Although it was a relatively straightforward matter to take sightings of the stars to determine latitude, the angular distance from the Earth's equator, determining longitude, the angular distance east or west of a set reference position, was more problematic. Even for latitude there had to be a reliable catalogue of the bright stars to help fix one's position. But for longitude it was all a matter of knowing the time difference between one's actual position at sea and the reference position. Since the Earth rotates 360 degrees in 24 hours, then 1 hour of time difference corresponds to 15 degrees' difference in longitude. And this in turn means that in order to be accurate to within 1 degree of longitude requires a clock accurate to within 4 minutes. Finally, since the circumference of the earth is 24,900 miles at the equator, a ship sailing there, off by just one degree of longitude, could be 70 miles from its destination.

Setting the local time of a clock at sea was not a problem; noon was simply the time when the Sun reached its highest elevation in the sky each day. But how certain could one be about the time back at the reference position? The clocks of the day, although precise enough while on firm land, could not hold their time with accuracy during a rough sea journey.

Instead of using a clock a mariner had to rely on the "course made good," the distance believed to have been traveled through the water, to provide a crude estimate of longitude. The distance traveled was worked out from the speed with which the ship passed an object thrown overboard—which is how the term "knots" originated, from the number of knots tied in a line that ran out while the ship moved. But the method was extremely imprecise, and the estimation of longitude remained largely a matter of luck and prayer. Ships continued to be lost at sea because navigation techniques remained so uncertain. It was believed, at least by the scientists of the day, including Newton, Hooke, and Flamsteed, that the secret of the accurate determination of longitude must lie in the heavens.

The Royal Society had long urged King Charles II to build an observatory dedicated to the longitude problem. Initially, the proposal did not catch the king's imagination. But, the king had taken as his mistress Louise de Quérouaille, a dark-haired beauty from Brittany in France, and made her the Duchess of Portsmouth. (His courtiers always suspected her of being a spy for French rebels, but this was never proved.) The duchess had a mysterious friend from Brittany, an amateur astronomer named Le Sieur de Saint Pierre, who claimed that he could solve the longitude problem using the Moon, and the duchess secured him a position in the royal court.

The king arranged for a Royal Commission to investigate Saint Pierre's claim. The commission included Robert Hooke, who was soon to become secretary of the Royal Society, and Christopher Wren, who in 1681 would become the society's president. (As well as being a successful architect, Wren was an outstanding mathematician and keen astronomer.) But the commission needed advice from an experienced observational astronomer, and John Flamsteed's help was sought.

It did not take Flamsteed long to demonstrate that Saint Pierre's method was inherently inaccurate. In questioning the Frenchman, Flamsteed established that his knowledge of astronomy was rudimentary at best. The report of the Royal Commission, based largely on Flamsteed's input, argued that there would be no easy route to the determination of longitude. More accurate data on the locations of stars and the movement of the planets and the Moon were needed. The Royal Society had been right. A new observatory dedicated to the accurate determination of the position of the stars was urgently needed to address the longitude problem. This time the king reacted, with the strong encouragement of a trusted courtier, Sir Jonas Moore. (The fact that France already had a royal observatory may have helped sway the king.) Moore had been the personal tutor to Charles's brother before the civil war. Following the restoration of the monarchy, Moore had returned to court as a favorite of the new king. He was made Master of the Ordinance, with control over the weaponry, battlements, and maps needed for the defense of the nation. His was a position of great power and influence. But Moore was also a learned man, with a love of mathematics and astronomy. The young John Flamsteed had come to his attention, and he was quick to recognize his ability. Moore acted as a powerful patron for the young astronomer in his early years.

On June 22, 1675, King Charles II granted a warrant authorizing the construction of the Royal Observatory for the purpose of "the finding out of the Longitude of places for perfecting Navigation and Astronomy." The king provided the location for the observatory in the royal park at Greenwich and the funding to build it (a sum of £520 was raised from the sale of old gunpowder). With the strong encouragement of Jonas Moore, he appointed John Flamsteed as his "Astronomical Observator." The twenty-eight-year-old clergyman's royal instructions were clear:

> Whereas, we have appointed our trusty and well-beloved John Flamsteed, master of arts, our astronomical observator, forthwith to apply himself with the most exact care and diligence to the rectifying the tables of the motions of the heavens, and the places of the fixed stars, so as to find out the so-much desired longitude of places for the perfecting the art of navigation.

The king commissioned Wren to design the new observatory "for the observator's habitation, and a little for pomp," as Wren put it. Hooke also assisted Wren in the design of the observatory (the two men had played a key role in the redesign and rebuilding of London following the Great Fire of 1666). However, because the funding available to Wren was limited, he was forced to use the existing foundations of an old castle watchtower, called Humphrey's Tower. (Humphrey, Duke of Gloucester, was the brother of King Henry V. King Henry VIII had used the castle at Greenwich as a hunting lodge.) Since the position of the observatory at Greenwich was later to define the prime meridian for specifying the longitude, it is amusing to recall that its exact location was entirely a consequence of building convenience. Secondhand bricks from Tilbury Fort and wood, iron, and lead from a demolished wing of the Tower of London also helped Wren to keep costs to a minimum.

Flamsteed laid the first stone for the new observatory at 3:14 P.M. on August 10, 1675. He prepared a horoscope for the observatory based on this time and date. It was intended in jest, since he certainly did not believe in astrology, and on the horoscope he scribbled, *"Risum teneatis amici"* ("This will keep you laughing, my friends").

During construction of the observatory, Flamsteed was provided with accommodation in one of the houses built into the Tower of London, and he carried out astronomical observations there. Flamsteed complained that the ravens at the tower disrupted his work by perching on his instruments and fouling them. The king was on the point of ordering the ravens to be disposed of, until he was reminded by courtiers of the tradition that when the last raven left the tower it would fall—and the monarchy with it. The ravens stayed. Until the observatory was completed Flamsteed was offered alternative, raven-free accommodation at the Queen's House at Greenwich from where he could easily oversee the construction of the observatory up the hill. Meanwhile, Saint Pierre returned to Brittany, a discredited figure, and the Duchess of Portsmouth didn't retain the king's favor much longer. The king soon had a new mistress. Intriguingly, he made her the Countess of Greenwich.

At this time the most accurate catalogue of star positions available was that which had been produced from observations by the Danish astronomer Tycho Brahe some eighty years earlier. Tycho's observations had been made with the unaided eye, so his star positions were less accurate than the positions later instruments with telescopic eyepieces would achieve. Tycho died before the publication of his results, but his pupil Johannes Kepler published them in 1627. Kepler was best known for developing laws describing the elliptical orbits of the planets about the Sun. Flamsteed was a great admirer of Tycho and referred to him as "the noble Dane" and "the greatest prince amongst astronomers." In setting up his observatory and its style of operation, Flamsteed had copied many of Tycho's observing methods. He even got Jonas Moore to commission engravings of the observatory and its instruments after the style of the etchings in Tycho's monumental work, the *Mechanica*. A young artist called Francis Place produced twelve etchings, which captured beautifully the atmosphere of the nascent observatory at Greenwich.

Although Flamsteed's official title was "Astronomical Observator," he styled himself as "Mathematicus Regius"—the same title enjoyed by Tycho when he held an imperial posting in Prague. Before long people started referring to him as the "Astronomer Royal." While Newton was reshaping science by describing natural phenomena in theoretical terms, Flamsteed followed Tycho's lead in generating a sound base of observations rather than pursuing theoretical goals. Flamsteed's commitment to careful observation throughout his time as Astronomer Royal followed clearly the path defined by the noble Dane.

Flamsteed's salary was a relatively modest £100 a year, from which he had to pay £10 tax and fund his own instruments. Jonas Moore provided some funding for equipment, but this source was lost when Moore died in 1679. Flamsteed found survival difficult, and he wrote to his government paymasters:

> Starve me not out, for my allowance you know is but
> small and now they are three quarters in my debt I fear
> I must come down into the country to seek some poor
> vicarage, and then farewell one experiment.

He did indeed have to seek some poor vicarage and tutor students for additional income—but he didn't starve, and he did eventually complete his life's commission, despite overwork, ill health, inadequate funding, and a great deal of conflict with Isaac Newton.

The Mesmerizing Moon

Although Flamsteed promoted the importance of astronomical observations to solve the longitude problem, he quickly realized the hopelessness of this position after commencing his observational program in 1675. What accurate astronomical timekeepers were there? Surely the Moon was an obvious candidate. If one could predict with precision the motion of the Moon across the heavens, then this could be used to tell the time at any location just as accurately as the motion of the hand of a clock across its numbered dial. However, the behavior of the Moon could not yet be predicted with sufficient accuracy using the theories then available.

The challenge was, in fact, threefold. The first challenge was to define the positions of the stars used as reference points (analogous to knowing exactly the positions of the numbers on the face of a clock). The second challenge was to understand the apparently erratic motion of the Moon across the heavens (analogous to understanding the motion of the hour hand of a clock). Finally, there had to be an instrument able to measure the angular distances between the Moon and suitable reference stars with precision and under the worst of conditions experienced at sea. (This final challenge was analogous to interpolating the position of the hour hand between the numbers on the face of a clock to determine time in minutes and seconds). None of these three challenges was trivial. This proposed method for determining longitude was called the "lunar distance" method—"lunars," for short.

Flamsteed turned to the moons of Jupiter as a possible answer to the difficulty of the "lunars," and in 1683 he published an article supporting their use for navigation. Galileo had discovered Jupiter's moons when he first used a telescope for astronomical purposes, and it seemed that they followed precisely the laws for planetary motion developed by Johannes Kepler. But there was a practical problem. How could one reasonably set up a telescope able to track the moons of Jupiter and make precision observations on the deck of a tossing ship? In addition, the exact time a moon slipped behind Jupiter was often difficult to determine. Despite its theoretical attractions, the idea lacked practical credibility. Nevertheless, Flamsteed maintained an interest in using Jupiter's moons to determine longitude, at least on land, and some of his later correspondence with Stephen Gray was on this topic. But for navigation at sea, it was back to the Moon.

Flamsteed started using a theory for the Moon's orbit developed fifty years earlier by an English clergyman named Jeremiah Horrocks. Horrocks had died at just twenty-four years of age, yet in his short life he had made significant contributions to astronomy. He applied Kepler's laws of planetary motion to the Moon, describing an elliptical orbit influenced by both the Earth and the Sun. In its original form this theory was not sufficiently accurate for use in navigation, but Flamsteed introduced various improvements to it. He made all time measurements relative to "Greenwich Mean Time" (his innovation for defining a reference system for time), and he determined that the Earth revolved on its axis at a uniform rate, rather than with a seasonally varying rate—a fiction that had persisted since Kepler's time. Flamsteed published his *Doctrine of the Sphere* in 1681, containing his version of the lunar theory. But still this revised theory did not fit well enough with observation, and the accurate prediction of the motion of the Moon across the heavens remained as problematic as ever. At the heart of the problem was an understanding of the orbits of astronomical bodies, and spectacular progress was being made on that front.

The scientists of London often met at fashionable London coffeehouses to discuss philosophical matters. On one such social occasion involving Robert Hooke, Christopher Wren, and Edmond Halley in January 1684, the topic of conversation had turned to the nature of gravity. Halley was a talented astronomer, perhaps best known for his observations of comets and predicting their periodic return. When Halley suggested that the elliptical orbit of a planet around the Sun might be explained by gravitational attraction varying as the inverse of the square of the distance from the Sun to the planet, Hooke boasted that he had already developed a mathematical proof demonstrating just this fact. Despite the encouragement of his colleagues, however, he was never able to produce his claimed proof.

Halley was much intrigued by this idea and persisted in his search for a mathematical formulation. Thus the following summer he traveled to Cambridge to meet with Newton. Despite Newton's isolation in Cambridge, it was known that he had developed new ideas about the nature of gravity. Halley wished to ask Newton whether he could describe the path of a body subject to an inverse square law for gravitation. He was unprepared for Newton's immediate response, that it would be an ellipse. What is more, Newton already had derived a mathematical proof while at Woolsthorpe escaping the plague. The sincerity and enthusiasm of the young man impressed Newton, and a few months later he sent Halley a treatise entitled *"De Motu Corporum in Gyrum"* ("On the Motion of Revolving Bodies"). There, revealed for the first time, was the mathematical proof of the inverse square law of gravity explaining the elliptical orbits of the planets. Halley persuaded Newton to publish the results of his research on gravity and the nature of forces. Indeed he offered not only to edit the work, but also to pay for its publication. The result of Halley's encouragement was the monumental *Philosophiae Naturalis Principia Mathematica*, published in 1687. Since the *Principia* established Newton as the greatest scientist of the age, the bond between Newton and Halley was understandably a strong one.

Although Newton and Halley were close friends, the same could not be said of Halley and Flamsteed. Flamsteed had fallen out with Halley a decade earlier. They had once been friends, but friendship turned to hostility some time between 1676 and 1678. Flamsteed's hatred for Halley was based initially on the latter's libertine lifestyle rather than on any scientific issue. Halley's heavy drinking and blasphemy could not be condoned by a man of the church. Lifestyle differences had soon manifested themselves as scientific differences.

If anyone was going to develop a sound theory for the motion of the Moon, that person was likely to be Isaac Newton. His *Principia* had redefined science. The nature of gravity seemed to have been determined, and the forces behind the physical universe appeared to have been explained, thanks to Newton's genius. If the planetary orbits could fall under Newton's spell, then surely a theory for the Moon's motion could be perfected by the greatest genius of the age? The problem was a complex one—gravity held the Moon in orbit around the Earth, which in turn was held by gravity in an orbit about the Sun. But surely the "intellectual master of the universe" could solve this three-body problem?

Why was it so difficult? Just as the Sun has a strong gravitational influence on the Earth, so too it affects the Moon and introduces many complex motions into the Moon's orbit. Thus the theory of the Moon's motion was not going to succumb simply to the clear guidance provided by the *Principia*.

In 1694, Isaac Newton (now aged fifty-two and still at Cambridge) and a brilliant young mathematician colleague,

David Gregory, visited Flamsteed at the Royal Observatory at Greenwich. Gregory had come under Newton's influence on first reading the *Principia* and became one of Newton's most staunch supporters. The meeting at Greenwich was not an entirely straightforward affair. Gregory had recently published details of a theory for the Moon's motion based on Newtonian mechanics, and in this work he had used observations supplied by Flamsteed to Newton. Flamsteed had received a written assurance from Newton that he would not pass any of his observations to anyone else without Flamsteed's agreement. He was thus not pleased to find Gregory using his results without any form of acknowledgment that they had come from Greenwich. Newton had already demonstrated that he could take the ideas and results of others without acknowledgment; yet he himself would be driven into fits of uncontrolled rage if his own monumental contributions were not fully acknowledged. Newton—and others—seemed to believe that since Flamsteed's position as Astronomer Royal was essentially a public appointment, his observations should be made freely available to be used by anyone without any form of acknowledgment. However, Flamsteed also had his pride. He was described as being possessed of an uncompromising attitude—an intemperate man even by the standards of an intemperate age.

Despite a certain feeling of irritation, Flamsteed did show his visitors a listing of his observed positions for the

Moon compared with the positions predicted by his revised Horroxian theory. The difference between theory and observation could be up to 12 minutes of arc—there are 60 minutes of arc to a degree. Since each arc minute in lunar position translates to 27 arc minutes in longitude, a 12-arc-minute error in lunar position was equivalent to a 5.5-degree error in longitude. Such errors were unacceptable—an error significantly less than just 1 degree was needed for navigation purposes. As far as solving the longitude problem was concerned, the lunar method remained too inaccurate to be of any possible use.

For Newton this discrepancy between theory and observation only acted as further stimulus to understand the apparently erratic motions of the Moon. He saw this task not only as a further verification for his theory of gravitation but also as the means of solving the longitude problem. He left Greenwich that day in 1694 recognizing that here was a new challenge to his supreme intellect.

Following this meeting there was a steady flow of letters between Newton and Flamsteed on the lunar problem. The early letters were highly courteous and helpful on both sides. Newton saw his own role in the lunar theory as paramount, and he made no pretense at modesty:

> For I find this theory so very intricate, and the theory of gravity so necessary to it, that I am satisfied it will never be perfected but by somebody who understands the theory of gravity as well, or better than I do.

Through the winter of 1694–1695, the letters demonstrate an apparent mutual respect and amity, but unfortunately that proved to be short-lived. Newton tried to resolve the difference between predicted and actual positions of the Moon by direct application of his theory of gravity as described in the *Principia*. It was not, however, going to be as simple as he had hoped.

At first Newton used an uncharacteristic flattery to encourage Flamsteed to provide him with a continuous flow of lunar data:

> And for my part I am of the opinion that your observations to come abroad thus with a theory which you ushered into the world and which by their means has been made exact would be much more for their advantage and your reputation than to keep them private until you die or publish them without such a theory to recommend them. For such a theory will be a demonstration of their exactness and make you readily acknowledged the most exact observer that has hitherto appeared in the world. But if you publish them without such a theory to recommend them, they will only be thrown into the heap of the observations of former astronomers until somebody shall arise that by perfecting the theory of the moon shall discover your observations to be more exact than the rest.

Newton's encouragement was clear: provide him with precise lunar observations so that he, Newton, could produce the perfect theory of the Moon's motion—the longitude problem could thus be solved, and Flamsteed's place in history would be secured (albeit in a subservient role to the great Newton).

Flamsteed initially obliged by passing on to Newton everything that was asked of him. However Newton, who expected perfection in himself, was intolerant when some errors crept into a few of the observations. It was of no account that Flamsteed blamed the errors on his assistants. Newton felt that Flamsteed should not have passed the data on to him until he had checked them personally. Even while criticizing the data, Newton did not let up on his demands for more. And his initial optimism about determining a theory for the Moon's motion turned to frustration as the problem eluded solution.

Flamsteed's inability to meet all of Newton's demands led to an unfortunate change in tactics by Newton. Hinting that Flamsteed was too lazy to provide all the reliable data needed, Newton offered to pay him for it. Flamsteed, a proud man of deep principle, was profoundly upset. He replied:

> All the return I can allow or ever expected from such persons with whom I corresponded is only to have the result of their studies imparted as freely as I afford them the effect of my pains.

Flamsteed undoubtedly held Newton in the highest regard scientifically at this time and was desperate to stay on good terms with him. He clearly wanted to provide Newton with observations because of the reflected glory this would bring. But Newton apparently saw Flamsteed as little more than the gatherer of routine data, merely a technician. For his part Flamsteed saw himself as the preserver of the proud tradition in observational astronomy established by "the noble Dane" Tycho Brahe. Here was a classic case of a clash of two great egos, each man convinced that he was pursuing a legitimate path toward understanding.

Desperate to reestablish his credibility with Newton, Flamsteed attempted to make his own calculations based on his lunar observations, and he sent them to Newton. Again errors had been made. Newton said that he did not want the calculations—only the original observations:

> If you like this proposal, then pray send me your first observations for the year 1692 and I will get them calculated and send you a copy of the calculated places. But if you like it not, then I desire you would propose some other practicable method of supplying me with observations or else let me know plainly that I must be content to lose all the time and pains hitherto taken over the Moon's theory.

The politeness and mutual respect evident in the early part of the exchange of letters had broken down into barely disguised frustration and bitterness. In his private notes Flamsteed described Newton as "hasty, artificial, unkind, arrogant." There can be no doubt that he was all of these; yet it is obvious from other writings that Flamsteed did recognize Newton's true genius and held him in grudging professional esteem.

However Flamsteed felt about him, Newton could not accept that any problem might be beyond his intellect to solve, and he made it clear to his friends that the cause of his failure to deliver a theory of the Moon's motion was John Flamsteed. His acolyte David Gregory had written after the breakdown in collaboration in 1696:

> On account of Flamsteed's irascibility the theory of the Moon will not be brought to a conclusion, nor will there be any mention of Flamsteed.

But access to reliable data was not the reason for Newton's failure to solve the problem. The reality is that Flamsteed provided Newton with an impressive amount of lunar data, of higher accuracy than had hitherto been obtained; it was simply not correct to claim that Flamsteed provided inadequate observations or deliberately withheld information at this stage. The real problem was that the theory was extremely complicated and was not going to be easy to determine without a better intuitive feel for the complexities of the Moon's motion than either Newton or Flamsteed possessed at the time.

Newton's genius was soured by his strong intolerance and temper, and his legitimate scientific advances were confused by his meddling with alchemy, astrology, and religious extremism. No one would doubt his massive contributions to science; however, some of his writings revealed the confusions of a troubled mind. The bitter rancor he unleashed on those who questioned his ideas could not be considered the behavior of someone of balanced temperament.

Moon to Mint

In 1695 Newton temporarily abandoned the lunar problem. Charles Montague, who had first met Newton when he entered Trinity College as an undergraduate some sixteen years earlier, was now rising fast through the political system. And such was Newton's reputation that Montague could see political advantage in associating himself with the great scientist. Montague had promised to obtain Newton an influential and highly paid position in London. This took longer than Newton had anticipated, and Newton began to doubt Montague's sincerity. However Montague was, eventually, as good as his word, and in March of 1695 he wrote to Newton offering him the post of Warden of the Royal Mint:

> I am very glad that at last I can give you a good proof of my friendship, and the esteem the King has of your merits. Mr Overton the Warden of the Mint is to be made one of the Commanders of Customs, and the King has

promised me to make Mr Newton Warden of the Mint.
The Office is the most proper for you; 'tis the Chief
Officer in the Mint, 'tis worth five or six hundred
pounds per annum, and has not too much business to
require more attendance than you may spare.

Although the Warden was essentially second in command, the post was prestigious enough and provided a good salary. Moreover, it was a step toward the top job, Master of the Mint, which Newton would secure a few years later. Certainly there was much to be done. The nation had lost control of its coinage, and inflation was rife; "clippers" were removing the outer edges of the gold and silver coins, and counterfeiting was easy. It was even known that workers at the mint were profiteering. Five hundred years earlier they had been taught a brutal lesson when, invited to Winchester to celebrate Christmas, they were castrated and had their right hands cut off for dishonesty and as a warning to future workers. But memory of this drastic historic warning had waned, and dishonesty within the mint was again rife. Now Newton would bring his methodical reasoning and organizational skills to the problem of counterfeiting and dishonesty. New coins were produced to replace the old, with milled edges containing the engraving *Decus et tutamen* ("An ornament and a safeguard") that could immediately reveal any illegal clipping.

As Newton planned his move to the mint, he broke off correspondence with Flamsteed with a terse explanation:

I have not yet got any time to think of the theory of the
Moon nor shall I have leisure for it this month or above:
which I thought fit to give you notice of that you may
not wonder at my silence.

Newton's sense of frustration at not having been able
to solve the lunar problem was intense. But Flamsteed felt
that all that he had done to help Newton should not go
unacknowledged. Flamsteed wanted to make public that
he had spent over a year obtaining precise sets of lunar
measurements for Newton, so that it would be understood
that it was not a lack of willingness on his part to provide
the observations that had led to failure. Newton would
have none of it, writing to Flamsteed:

I was concerned to be publicly brought upon the stage
about what, perhaps, will never be fitted for the public,
and thereby the world put into an expectation of what,
perhaps, they are never likely to have.

For the time being, as far as interactions between
Newton and Flamsteed were concerned, that was that.
However, many of the frustrations Newton felt in dealing
with Flamsteed on the lunar theory were to carry forward
to the future major conflict with the Astronomer Royal,
which would embrace even his friends.

But for now Newton had nothing more to do with
Flamsteed. He was on his way to London where long-term

residents, such as Wren, Hooke, and Halley, dominated science. It did not take long for him to make his presence felt.

New Moon

By the time Newton assumed the presidency of the Royal Society in 1703, he was busy positioning himself for an enhanced place in history by ensuring that all his research results from decades earlier were now published. Several scientists were doing research in his name. Such was the magic of the Newton connection that there were many young researchers keen to extend his ideas and share the credit for this new work with the great man. Newton's scientific reach was impressive, and he continued to control the progress of research in Cambridge, Oxford, and London through orchestrating the placement of his friends in key academic positions. David Gregory was appointed Savilian Professor of Astronomy at Oxford at Newton's suggestion. Roger Cotes, a young Trinity graduate, was positioned in the new Plumian Professorship of Astronomy at Cambridge with Newton's encouragement. Edward Paget, a fellow of Trinity, was made Master of the Mathematical School of Christ's Hospital in London. The Newtonians rapidly gained control of British science in the early years of the eighteenth century.

In 1702 Gregory published a textbook on Newtonian astronomy, titled *Astronomiae Physicae*. It contained a five-page section on Newton's *A Theory of the Moon's Motion.* Was this a result of Newton's earlier work of 1694–1695, passed on to Gregory at that time and only now appearing in print? It seems unlikely, since in 1695 Newton was making no secret of his frustration at not being able to solve the lunar problem. It is more likely that Newton had finally produced a draft of a solution he found acceptable much closer in time to the publication of Gregory's book. Gregory said that Newton wrote the manuscript of *A Theory of the Moon's Motion* in February 1700. If this date is accurate, then it demonstrates that Newton's youthful ability to take on many tasks and assume an impressive workload had not diminished with age since in the very same month Newton took over the position of Master of the Royal Mint. He was now in charge. On February 3 a royal edict noted:

> Know you that we for diverse good causes do give and
> grant unto our trusty and well beloved subject Isaac
> Newton Esq, the office of Master and Worker of all our
> moneys both gold and silver within our Mint in our
> Tower of London, and elsewhere in our Kingdom of
> England. And know you that we for the considerations
> aforesaid have given and granted, and by these presents
> do give and grant unto the said Isaac Newton all
> edifices, buildings, gardens, and other fees, allowances,
> profits, privileges, franchises and immunities belonging
> to the aforesaid Office.

Gregory claimed that Newton's *A Theory of the Moon's Motion* produced lunar position predictions with an accuracy better than 2 arc minutes, offering terrestrial longitude estimates equal to the much-sought-after 1 degree. Flamsteed merely dismissed Gregory as a "closet astronomer" (since Gregory made no observations himself) and said that the theory gave no real improvement over existing tables. Edmond Halley acclaimed the new theory, saying that it greatly improved lunar predictions. Indeed, there was no doubt that Newton's *A Theory of the Moon's Motion* was a very significant advance over Flamsteed's version of the Horroxian theory. However, the method was essentially empirical, refining earlier equations so that they provided improved predictions. It was still lacking the sublime physical insight one would have expected from Newton and suggests that the solution may have owed more to the impressive mathematical skills of David Gregory. The French historian Jean Bailly noted in 1779:

> He has left the Horroxian theory as a semblance, to
> await the true theory and take its place.

Bailly noted skeptically its lack of connection to the law of gravitation:

> It does not have any physical cause.

Newton himself was not satisfied with this theory, since it still lacked a physical explanation in terms of his theory of gravitation. He needed more and better lunar data if a full description of the Moon's behavior was to be made. Hence, once he had the authority that the presidency of the Royal Society endowed, he hatched his plan. He would get royal support to force Flamsteed to publish all his observations, among which Newton believed lay a wealth of extra lunar data.

But if Newton had thought his plan to call the Astronomer Royal to account was foolproof, he had not counted on Flamsteed's pride or his intransigence. Although Flamsteed had earlier shared his data more than Newton was willing to believe, by 1705, when he had begun to prepare the publication of his observations in earnest, he used every conceivable delaying tactic to keep his precious observations from Newton's reach until he deemed them ready. The battle over the Greenwich astronomical catalogue would prove to be a long one.

Still, Newton, the great schemer, had his other line of attack on Flamsteed: promoting a rival observatory to Greenwich at his old university, Cambridge. The center of immediate interest was to move to an unexpected Cambridge connection, and Stephen Gray found himself surprisingly tied up in Newton's malicious designs.

CHAPTER 3

The Cambridge Connection

[Stephen Gray] was the father, at least the first propagator, of electricity.

—WILLIAM STUKELEY

As Stephen Gray walked through the magnificent gatehouse entrance to Trinity College, Cambridge, he felt as if he were in a dream. Even for someone used to the size and beauty of Canterbury Cathedral, the Trinity courtyard was an imposing sight. It seemed as if he were on a pilgrimage to a hallowed shrine of learning—just as those who visited his native Canterbury were often on a pilgrimage to a hallowed shrine of Christianity. He thought how proud his father and mother would have felt that one of their children had taken employment in this revered center of scholarship. And how proud his brother Mathias would have felt. Mathias had died the previous year, 1706. His death had been a further devastating loss to Gray. There had been so many untimely deaths among his nearest and dearest. Mathias, with his success in business and as mayor of Canterbury, had brought so much credit to the Gray family. Canterbury now seemed a sad and forlorn place for Stephen Gray. Dyeing had become a burdensome occupation, affecting his health and peace of mind. At age forty, his life seemed to be going nowhere, until he received an unexpected offer of a job as an assistant at a new astronomical observatory being built at Cambridge.

The Pilgrim's Way

The trip from Canterbury to Cambridge had taken two days. The first stage of the journey, by coach to London, would have allowed a visit to John Flamsteed at Greenwich. Gray, surprisingly, let the opportunity pass. Flamsteed had counseled him not to go to Cambridge, and he felt it would now cause further upset if he were to visit Greenwich en route to a job Flamsteed had advised him so firmly against taking. After an overnight stay in a London tavern, an uncomfortable coach journey on the following day completed the trip to Cambridge. His nephew, John, who was studying medicine at Cambridge, met him on arrival.

It was John Gray who had organized the appointment at Trinity for Stephen. He had made friends with a Trinity fellow named Roger Cotes, who was setting up the new astronomical observatory at Cambridge and needed to employ an assistant. Knowing of his uncle's love of astronomy and considerable experience as an observer, John Gray pressed Roger Cotes to employ him.

The Trinity porter escorted uncle and nephew across the mighty central courtyard of the great college and up a winding staircase to the loft chambers that would be Stephen Gray's new home. The accommodation was spartan. Only a bed, a pottery washbowl, and a chamber pot outfitted the small bedroom. An even smaller study contained a desk and single chair. But for Gray it all seemed perfect. He did not worry that he had to walk down three flights of narrow stairs to fetch water or that there was no fireplace in his room to provide heat. Being at Trinity College was a source of such enormous honor that a place in the stable hayloft would have sufficed.

The Blessed Trinity

King Henry VIII had established Trinity College in 1546. He amalgamated two earlier colleges, with the expressed intention of creating a new college at Cambridge that would rival in splendor Christ Church in Oxford (founded as Cardinal College by Cardinal Wolsey, Henry's discredited chancellor). Henry barely could have imagined the academic heights his new college would attain, fostering the scientific achievements of Isaac Newton. Although he had laid waste to the English monasteries, Henry was a strong supporter of the endowment of the Cambridge and Oxford universities. He told his courtiers:

> I tell you, Sirs, that I judge no land in England better
> bestowed than that which is given to our Universities.
> For by their maintenance our Realm shall be well
> governed when we be dead and rotten.

Cambridge and Oxford were the two great English medieval universities, and they have remained at the forefront of intellectual achievement ever since. Oxford is the older, and Cambridge in fact owes its existence to Oxford. In 1209 a band of scholars escaped rioting in Oxford by traveling to Cambridge (some eighty miles distant) and establishing themselves there. The two universities were based from the outset on a confederation of independent colleges. The early colleges of both were richly endowed and were grand in their design. Their typical cloistered designs reflected those of the monastic houses that formed the earliest centers of scholarship.

The town of Cambridge to which Stephen Gray moved in 1707 had a population of some seven thousand; three thousand of these were students and their tutors. The university dominated life in Cambridge. Many of Cambridge's inhabitants beyond the university, such as tavern owners, prostitutes, and vagabonds, made their living from preying on the frailties of the students. The town was a threatening place, with thieves seeing wealthy students as an easy source of illicit income. The murder of a student during a violent robbery was, sadly, a not infrequent occurrence. Yet in spite of the many uncertainties of life in the town, the university flourished as a haven of scholarship and learning.

Trinity College's reputation had been greatly enhanced through the achievements of Isaac Newton. It now prospered under an entrepreneurial master, Richard Bentley. The small lean-to shed against the chapel wall that had formed Newton's laboratory for his alchemy experiments still stood and could be easily sighted from the dormer window of Gray's chamber. The Newton "aura" still permeated Trinity a decade after his departure. The academic life of Cambridge had appealed to Newton from the day he entered Trinity College in the summer of 1661. As an undergraduate he had ignored much of the established curriculum (based on Latin, Greek and Hebrew, scripture and rhetoric), concentrating instead on the latest developments in mathematics and natural philosophy. His original intention had been to concentrate on chemistry. However, on a visit to the Stourbridge Fair a few months into his studies he purchased a book on astrology. Newton found that he could not understand his newly purchased astrology book because of the geometrical and trigonometric calculations it contained. He therefore purchased books on mathematics, and although he had never had any formal education in mathematics, he discovered that its logical structure and methods came naturally to him. He found the teachings of Euclid trivial, but he was excited by the writings of Descartes. He read widely in natural philosophy, theology, mathematics, and (surprisingly) alchemy.

Alchemy, the attempt to convert base metals to gold, was very much on the fringe of legitimate investigation by this time, its heyday having been in earlier centuries. Newton kept his interests in the subject a secret from his academic colleagues, although he did communicate his views through a clandestine network of fellow alchemists. In later life, when his scientific reputation was secure, he took extreme measures to disguise his excursions as a young man into the cult pseudoscience. Nevertheless, it does appear that Newton maintained a lingering interest in alchemy and the occult throughout his life, despite the fundamental contradiction between alchemy and the scientific doctrines he expounded publicly.

Newton's theological beliefs were just as clandestine and at odds with contemporary orthodoxy as his alchemy. He was an Arian—someone who denies the doctrine of the Trinity—and almost all of his early writings were on theological matters. All of this was kept well hidden, however, because of legal constraints upon him as an academic. Part of the mechanism of the Anglican confessional state of the era was that entrants to Oxford University and graduates of Cambridge University had to sign the Thirty-nine Articles of the Anglican Church, which affirmed the authority of the monarch and the doctrine of the Trinity. In fact, on four different occasions during his time at Cambridge, as he climbed from his B.A. degree to his M.A. to fellowship of Trinity College and then eventually to professorship, Newton would have had to have signed the Thirty-nine Articles. Had the controversial nature of Newton's theological beliefs been widely known, it is almost certain that his rise to power would have been thwarted prematurely by the university authorities. Other Arians who did make their beliefs widely known were removed from academic appointments. Many who found it impossible to sign the Thirty-nine Articles moved to Holland and elsewhere in Europe where alternative religious views were tolerated.

When Newton returned to the university in 1667 after it reopened following the plague, Trinity College elected him to a fellowship. He kept secret much of his research at this time, although his outstanding abilities were recognized. When Isaac Barrow, the then Lucasian Professor of Mathematics, was offered the post of chaplain to King Charles II in 1669, he nominated Newton as his successor. At just twenty-seven years of age Newton became the holder of a prestigious academic appointment, a post he formally retained until 1701, some five years after he eventually left the university for London. Normally a Cambridge professorship required its holder to take holy orders; Newton somehow managed to escape this requirement.

Although Newton was the best known of Cambridge's scholars, there were many others of considerable distinction. Many of them crossed the Atlantic during the seventeenth century to help found the early colonies that would eventually form the United States. One of these was John Harvard, who bequeathed half of his considerable estate to a new college at "Newe Town." The college would be renamed Harvard in his memory, and the town became known as Cambridge in recognition of the contribution of John Harvard and his fellow graduates who helped lay the foundations of the new nation. One of the first graduates of Harvard was George Downing, who left the New World to make his reputation in the Old. Ironically, his wealth would eventually endow Downing College in Cambridge, England. And it was for Downing that the street in London was named, where "Number 10" was to become famous as the official home of the British prime minister.

Cotes, the Competitor

Stephen Gray's employer at Cambridge, Roger Cotes, was a favorite of Newton and a protégé of Richard Bentley. It is interesting to speculate as to why Roger Cotes selected Stephen Gray as an assistant, even if encouraged to do so by John Gray. It is true that Stephen Gray had established a reputation as a careful astronomical observer, but a humble dyer was not the most obvious of assistants to a brilliant young academic appointed to a prestigious new chair and charged with creating a new observatory to compete with the best. Might someone else then have influenced Cotes's decision to employ Gray? Certainly not Flamsteed, since he could not have contemplated an astronomical collaborator joining a rival team, nor is it likely that Cotes would have listened to him. It seems highly likely that Cotes would have sought Newton's opinion, however, since the great man was taking a keen interest in Cotes's plans and was supporting the new observatory financially. Newton would have been well aware that Gray was a friend and scientific associate of Flamsteed. Might he have been tempted to agree to Gray's appointment to the nascent observatory as an indirect way of wounding Flamsteed? It is merely speculation, and there is no written evidence to support it, but this scenario would certainly fit the circumstances of the time as well as Newton's pattern of behavior.

Roger Cotes had been elected to the new Plumian professorship of Astronomy and Natural Philosophy at Cambridge in January 1706. He had been born in 1682 and educated at Saint Paul's School, London (the same school attended by Edmond Halley). He then moved to Trinity College, where he first came under Newton's influence. Bentley wished to inject some new blood into Trinity, and Newton helped to secure the Plumian professorship appointment for Cotes at the very young age of twenty-four. With Newton having supported Cotes's appointment to his old college, there could be no doubting where

Cotes's subsequent loyalty would lie when the Newton-Flamsteed confrontation later reached its peak.

With Newton's encouragement, Cotes set out to build his new observatory at Cambridge. Richard Bentley promoted the cause of the planned observatory with characteristic enthusiasm. He helped Cotes obtain sponsorship for instruments, and Newton donated a clock for the accurate timing of observations. With his dispute with Flamsteed unresolved, Newton saw a rival observatory to Greenwich as an alternative way of getting high-quality astronomical data to improve his lunar and other theories.

Cotes's new observatory was to be located above the gatehouse at Trinity College. It was an imposing site, since the gatehouse built for Henry VIII was one of the college's most impressive architectural features. Cotes's initial plans for the observatory and its funding are revealed in a letter he sent to an uncle in February 1708:

> Honoured Uncle, I have lately been at London. I found your letter at Cambridge upon my return. The occasion of my going up there was partly to view a large brass sextant of 5 foot radius (that had been made for us and is now finished) before it should be sent down. Whilst I was in town Sir Isaac Newton gave orders for the making of a pendulum clock which he designs as a present to our new observatory. The sextant will cost the college 150 pounds, and I believe Sir Isaac's clock can cost him no less that 50 pounds.

The £50 cost of Newton's clock for the observatory should not be underestimated, bearing in mind that Flamsteed was receiving just £100 by way of an annual salary plus running costs for Greenwich. An accurate clock was an essential piece of technology for an astronomical observatory, since so many calculations depended on accurate timings.

Richard Bentley reported the changes at Trinity in 1710:

> The College Gatehouse was raised up and improved to a stately Astronomical Observatory, well stored with the best instruments in Europe.

Elsewhere he noted that the observatory was

> the most commodious building for that use in Christendom, and without charging the College, paid for by me and my friends.

Bentley's description was exaggerated; the instruments were more than rudimentary, but they were never brought to a standard to match even remotely those of Flamsteed. The completion of the observatory took much longer than expected and cost far more than Bentley had intended. The workmen frequently downed tools, since they were not being paid. The final cost was at least three times Cotes's initial optimistic estimate.

Electrical Diversions

While he was at Trinity, Gray was able to venture for the first time into experiments with electricity. His experiments, carried out by candlelight late at night in his chamber, and sometimes in his nephew's study, would eventually have a profound impact. William Stukeley, a noted antiquarian of his day and one of Newton's contemporary biographers, came into contact with Gray at Cambridge. In his memoirs he recorded:

> Mr Stephen Gray of Canterbury was in our University as an Assistant to Mr Cotes, Professor of Astronomy, for whom they built the observatory in Trinity College; a very ingenious man well versed in Philosophy, Astronomy, Optics, Mechanics, etc. . . . Uncle to Dr John Gray. . . . We three used to smoke many a late pipe together, and try various experiments in Philosophy.

Elsewhere Stukeley noted:

> Electricity was found out by Mr Stephen Gray of Kent. He was uncle to Dr Gray. Stephen Gray showed us his experiments about it at Bene't College where I was a student there in 1708.

Stukeley clearly felt that Gray's contribution to electricity had wrongly been neglected, since very much later, in 1752, when he was a prominent member of the Royal Society, he wrote:

> One purpose executed by the Royal Society is that of being a philosophical court of record, to register the memorials of discoveries in any of the branches of natural knowledge. For want of this kind of care we are sometimes detained in hearing complaints and grievances, some people being deprived of their just due in these respects. In order to this purpose of record I offer the ensuing account of electricity, that great and universal agent, which now entertains and employs all the chiefs in the philosophic world. In the year 1708, Stephen Gray, who then lived in Cambridge, often visited his nephew, Dr John Gray, at Bene't College with me. He showed us many times his electrical operations with a long glass tube. He had a particular knack of exciting this property by friction with his hand, and was the father, at least the first propagator, of electricity.

Despite this contemporary account from a member of the Royal Society, there are few books on the history of electricity that make any mention of Stephen Gray, and if mention is made it is brief and of meager consequence. The reason is that Gray's contemporaries largely ignored his work on electricity. Yet his discoveries did not go entirely unnoticed since some were purloined by a few individuals and presented as their own.

In January 1708 Gray wrote to the secretary of the Royal Society, by now the Irish physician Hans Sloane. The letter was headed "From my Chamber in Trinity College in Cambridge." The main purpose of the letter was to describe his new research in electricity, but he starts courteously by thanking Sloane for sending him copies of the *Philosophical Transactions*. He then pays tribute to the scientist Francis Hauksbee, curator at the society (the post initially occupied by Robert Hooke), for reporting the electricity experiments that had inspired Gray's own work. Newton had appointed Hauksbee to the curator post when he took up the presidency of the society.

The basics of electrical phenomena had been known since ancient times. Amber, rubbed with fur, attracts a pith ball. (The childhood trick of rubbing a comb to attract pieces of paper demonstrates the same effect.) The English scientist William Gilbert, at the end of the sixteenth century, was the first to use the terms "electric" (from the Greek word for amber), "electricity," "electric force," and "electric attraction." He was also the first to demonstrate that amber was not the only substance that attracts light objects when rubbed. By the early eighteenth century, two types of electricity had been identified (positive and negative) and electrical forces had been shown to be both attractive and repulsive, depending on the materials used. The rubbing of a glass tube was found to be a particularly effective way of generating electrical effects.

At this time, the accepted theory for electrical phenomena involved the creation of an electrical "virtue"; it was suggested that friction would remove what was referred to as the "humor" from a rubbed material and create an "effluvia" flowing from the electric object to influence nearby objects. The language was picturesque and the physical understanding negligible. If we substitute the modern terms "electric current" for "virtue," "electric charge" for "humor," and "electric field" for "effluvia," however, then the historical descriptions make a little more sense (although the effluvia theory could not be sustained as more discoveries about electrical phenomena were made). Hauksbee was acknowledged as being one of the most successful demonstrators of electrical phenomena at the time. From 1706 to 1711 he published in the *Philosophical Transactions* ten contributions on his electrical experiments. Stephen Gray was an avid reader of the *Philosophical Transactions* when he could get his hands on copies, and he gained many of his experimental ideas from perusing their pages.

Gray's letter from Trinity College to Hans Sloane began:

Learned Sir,

I thank you for the continuance of your *Philosophical Transactions* to me. Those that have any delight in the knowledge of nature cannot but esteem them, whatever some may say who had rather lay the deficiencies they pretend to find in them on your *Transactions* than on their own unphilosophical inclinations.

I perceive Mr Hauksbee goes on still to oblige the Philosophical World with his Ingenious inventions, Luciferous Experiments, and Noble Discoveries, amongst which I take those to be not the least that relate to the Production of Light and electricity by the Attrition of Glass. The strangeness of the Phenomena, together with the facility of the operation in these Experiments he made with a Glass tube, encouraged me to attempt the Pursuance for these following Experiments. Some of them are at least so very surprising and others of them seeming to Trace out the Extent and current of the Luminous and Electric Effluvia of Glass. I presume it may be not unacceptable to give some account of them.

The Glass tube made use of was about the size of that made use of by Mr Hauksbee, but instead of rubbing it with Paper as he directs I found it to succeed better with me when I rubbed with my bare hand only.

Many of the experiments Gray conducted described the behavior of a feather under the influence of a charged glass tube, from which he tried to infer the nature of the electrical effects produced by the tube. For example:

Experiment 4th. When the feather is come to the Glass, and thence Reflected, if you follow it with the Glass it will flee from it and will by no means be made to touch it till driven near to the next wall in the Room or some other solid object, by which it will be attracted and freely Return to the Glass again. . . . So I have sometimes carried the feather across the Room at the Distance of 5 or 6 inches without touching it, and could move it upwards, downwards, inclining or horizontally, in a line or circle, according to the motion of the Glass. And if when the feather was floating in the air I rubbed the Glass, the feather would remove farther from it, yet would answer to the motion of my hand by a vibrating motion not to be accounted for by the motion of the air.

It is not difficult to see why Stukeley and John Gray were so entertained by Stephen Gray's demonstrations of such novelties—or why Francis Hauksbee should wish later to demonstrate them to the Royal Society. (Hauksbee conveniently ignored the fact that Gray had been the first to demonstrate certain peculiarities of electricity, and he failed to acknowledge Gray's work when he drew on it in his subsequent publications in the *Philosophical Transactions*.)

Gray described an electrical spark, similar to that produced from glass by Hauksbee, as flowing from his finger to an electrified sample. He found that the effect, accompanied by an audible noise, could be generated from sulphur, sealing wax, and amber, as well as glass:

Experiment 7th: That light which proceeds from one's finger when held near the tube in the dark Proceeds from it in a Conical Stream whose vertex is at the finger, and the noise it makes seems to proceed from the striking of the Glass by the Effluvia in the rapid motion from the finger.

Gray found that the most spectacular example of a spark could be generated using a sharp pointed stick:

Experiment 8th: A small sharp stick emitted light all the time the Glass tube was rubbing at the Distance of more than a foot, increasing or Diminishing its light according to the intensity of the force used in rubbing it.

Gray confirmed Hauksbee's observation that "moisture is the Enemy to these experiments." A dry atmosphere was needed. Gray could not complete his demonstrations so effectively when there were more than two people in the room, "I suppose from the motion of the Effluvia being hindered by the steam of their bodies." He was intrigued to note that:

The Glass of my Watch was Electrical without any other friction than it Receives in my pocket accidentally.

Although some of Gray's observations on his Trinity electricity experiments are simplistic, his careful approach revealed many of the fundamentals of electrostatics. Had he been encouraged to pursue his electrical research at this time, and had his experiments been openly reported, then there is every prospect that the science of electricity would have made much more rapid progress than it actually did. It would have been only through repeated experiments that Gray could have expected to reveal the true nature of the electrical forces and their application. Sadly, Gray received none of the encouragement he might reasonably have expected.

Gray's letter to the Royal Society from Trinity College, reporting epoch-defining experiments in electricity, was never published in the *Philosophical Transactions*. It is worth asking why it was not published, since prior to Newton's presidency in 1703 the majority of Gray's letters to the Royal Society had been published. After Newton assumed the presidency, however, none were published until 1720, when a single paper was accepted, and none were published thereafter until after Newton's death seven years later.

Unfortunately, letters from Gray to the society on astronomy shortly after Newton assumed the presidency paid generous tributes to Flamsteed. The praise was effusive:

That most learned, accurate and judicious astronomer

That most accurate astronomer, my most honoured friend

This great astronomer

The same excellent astronomer

Obviously Newton would not have enjoyed reading such praise of an individual whom he had come to despise. It is almost certain that these letters would have irritated him greatly, and the glowing praise of Flamsteed might have been all that was needed for him to have set aside Gray's letters rather than have them published in the *Philosophical Transactions*. The president had the ultimate say on which papers were published in the *Philosophical Transactions*. And while Gray's letter from Trinity on electricity made no reference to Flamsteed, Newton might have ignored it merely because he now recognized Gray to be one of Flamsteed's key supporters. Newton may by now have also heard that Gray was passing critical comment on the Cambridge observatory plans.

Writing from the old college of the society's president must have raised Gray's expectations that his letters would be published. But his research into electricity went unknown until his letter from Trinity was eventually found in the archives of Hans Sloane. Although Gray returned to electricity experiments many years later, he never made further reference to his first experiments made at Trinity. Had Sloane not been so meticulous in keeping correspondence, all we would have known would have been from Stukeley's rather vague references.

The Royal Society Journal Book notes that in response to Gray's letter describing his experiments Hauksbee was asked to "try out what is new in them." Hauksbee, however, copied many of the electrical novelties and demonstrated them to later meetings of the Royal Society as if they had been his own discoveries. The most blatant example of Hauksbee's piracy was his reporting the "luminous effluvia of wax and sulfur" just a few months after the society had received Gray's letter announcing this discovery. Later in 1711 Hauksbee again tried to lay claim to Gray's discoveries:

> It may be remembered what success I had in producing light through Bodies, such as Sealing Wax, Pitch, and common Sulfur.

Could it have been Hauksbee, then, who orchestrated the nonpublication of the Trinity letter? It seems extremely unlikely, since curators really could not exercise such influence. (As curator, a much more influential Robert Hooke could not stop the publication of Newton's work on optics thirty years earlier—although he felt it drew strongly on his own work, without due acknowledgment, and he did criticize it.) Gray had freely and generously acknowledged Hauksbee as the inspiration for his experiments, and surely Hauksbee would have welcomed such handsome acknowledgment appearing in the *Philosophical Transactions*. Would Sloane have stopped the publication himself? Since he had published so many of Gray's letters before the Newton presidency and had sent Gray gratis copies of the *Philosophical Transactions* (including during his time in Cambridge), it is hard to believe that Sloane would have wished to suppress his work. Also,

Hans Sloane eventually succeeded Newton as president of the Royal Society, after which all of Gray's letters were again published. Gray was first recipient of the society's Copley Medal in 1731, and Sloane was one of the trustees of the Copley Trust who selected the medal winner. Sloane was clearly a strong supporter of Gray.

On the other hand, it is difficult to escape the conclusion that Gray's letter from Trinity was not published because of the direct intervention of Newton—or if not his direct intervention, then the action of society officials who were under his clear instructions to suppress the contributions from the supporters of Flamsteed. It is impossible to believe that Gray's electrical experiments could have been rejected for publication because they were competing for space in the *Philosophical Transactions* with more worthy contributions. The *Philosophical Transactions* at this time are filled with a miscellany of trivia from fellows and nonfellows alike. There is no "smoking gun" proving Newton's involvement in the suppression of Gray's letter, for example in the form of a written instruction from him. But since almost all of Gray's letters to the Royal Society before Newton's presidency were published and all of his letters that followed Newton's death were published, the possibility that Newton stopped the publication of experiments from a friend and vociferous supporter of Flamsteed is not only possible but highly probable.

In addition to having his letters suppressed during Newton's presidency, Gray also discovered that the gratis copies of the *Philosophical Transactions* he had long received suddenly stopped about this time. And since he could not afford to purchase them, Gray had lost an important link to the real world of science. Gray seems to have been sufficiently disillusioned by the rejection of his 1708 letter that there is no evidence of his having returned to electrical experiments for more than a decade.

Had Gray's pioneering electricity results of 1708 received the public acknowledgment and encouragement they deserved, his monumental experiments of 1729 on the communication of electrical phenomena would likely have happened twenty years earlier. Moreover, if Newton had encouraged Gray's research, as he had that of Cotes, Gregory, Halley, and other acolytes, then the science of electricity could have been advanced by decades—perhaps more.

The Pilgrim's Return

Gray's partnership with Cotes at Trinity College was not a happy affair and did not last long. The cause of Gray's departure may have been a disagreement about his salary, but he was also unhappy that the Cambridge ambition was solely to rival the work of his friend John Flamsteed. In September 1708 Gray wrote to Flamsteed reporting that he was back in Canterbury, having made an unhappy departure from Cotes's employment. The dyeing business must have been left in trusted hands during the Cambridge sojourn. Certainly Gray assumed control of the business on his return to Canterbury. He wrote to Flamsteed:

Reverend Sir,

It having been a long time since I had the opportunity of writing to you, now that I have one I make bold to trouble you. I did not think you expected to hear from me from Cambridge, yet I often thought of you and should have been glad to have seen anything there worth your knowledge, but you will easily believe me if I tell you I saw nothing there that might deserve your notice. There was indeed that which they called their observatory, for no other reason that I could perceive than some time or other they intend to make it so. They have at length gotten one instrument, which when their observatory is built they may perhaps do something with in time. I need not describe the sextant because I am satisfied it is well enough known to you already. They talk of doing great things with it, but I dare say far short of what they expect. They propose to themselves no less than making a new catalogue of fixed stars, but when they come to see the difficulty of managing such an instrument I do not doubt they will be convinced of the vanity of such an attempt. But besides, as I have told them, the world is not to expect a more ample or more accurate catalogue than yours, whether we consider the goodness of your instruments, the method of your observations, or the long experience and indefatigable industry of the observer. Now that you have laid so good a foundation, I think they ought to build upon it, for if they pretend to lay one of their own I wish it may not prove to be a deceitful one. Their dealings with me give me some cause to fear it. I had better have taken your advice, which was more agreeable to my own inclinations, had I not been pre-engaged by the solicitations of my friends, but we little thought or suspected such men could have been so mercenary as I find they are. I need not go any further with the relation of the occasion of my leaving them because I believe it is known to you already.

Gray's unfavorable comparisons of the Cambridge plans with Flamsteed's work, while no doubt pleasing news for Flamsteed, must have been a source of tension with Cotes. Since Newton remained in regular contact with Cotes, Gray's criticisms are likely to have reached Newton, who was excessively protective of his favorites. The criticisms would have angered the great man.

It seems strange that Stephen Gray did not write to Flamsteed in the whole time he was at Cambridge, since he had written so freely before his time there and wrote even more frequently on his return to Canterbury. Since Flamsteed had urged Gray not to take the post, perhaps Gray had been too embarrassed to write. Certainly the letter on his return to Canterbury is almost apologetic in its tone.

Cotes, a very sociable young man, apparently let his socializing get in the way of professional discipline. Halley was to express frustration at Cotes's failure to produce a complete set of timed observations of the solar eclipse of 1715 because he was diverted by too many companions. Cotes died young, in 1716, at the age of just thirty-four years. On his passing, Newton paid tribute to his brilliant mind:

> Had Cotes lived, we might have known something.

Lacking Cotes's guiding hand, the observatory at Trinity, set up to compete with Greenwich, produced nothing of real astronomical value. None of Cotes's eighteenth-century successors to the Plumian chair did anything of note at the Trinity observatory, and the observatory fell into a state of disrepair. In 1792 the Plumian trustees noted sadly that

> the observatory and the instruments belonging to it were through disuse, neglect and want of repairs so much dilapidated as to be entirely unfit for the purposes intended.

The Trinity observatory was pulled down in 1797, having contributed very little to the understanding of astronomy. Gray's letter to Flamsteed on leaving Cambridge turned out to be strangely prophetic:

> They talk of doing great things with it, but I dare say far short of what they expect.

By contrast, Flamsteed's work at Greenwich would define new standards of excellence in observational astronomy. Thus the grand plan to have a Cambridge observatory challenging Greenwich had come to nothing.

Gray was disillusioned by his experience in Cambridge and by the nonpublication of his electrical discoveries. On his return to Canterbury, however, he renewed his astronomical observations with vigor. In the few years following his return there is a steady flow of letters to Flamsteed reporting sunspots, lunar eclipses, a solar eclipse, and several observations of the eclipses of Jupiter's moons for which there was a lingering interest in terms of the longitude problem. He made only one later written reference to his time at Cambridge.

By the time Gray returned to Canterbury, Flamsteed was embroiled in open conflict with Newton and Halley over the publication of his star catalogue—a confrontation that would lead in time to a "sacrifice to Heavenly truth." The battle between science's first knight and first Royal Astronomer could have no happy outcome and would set the cause of science back by many years.

CHAPTER 4

The Sacrifice to Heavenly Truth

How unworthily, nay treacherously, I am dealt with by Sir I Newton.

—JOHN FLAMSTEED

Robert Hooke died at Gresham College in March 1703, following twelve months of being bedridden and nearly blind. It was a sad and protracted end for a man whose intellectual vitality had carried a feeble and sickly frame through sixty-eight years. Just eight months after Hooke's passing, Newton was elected president of the Royal Society. So closely was Hooke associated with the culture and ethos of the society that Newton would never have agreed to become its president while his archenemy Robert Hooke was still alive. He had been the society's first curator, for a period its secretary, and one of its most prolific contributors. To a great extent the Royal Society was the type of organization it was because Robert Hooke had been the type of man he was. A proud, energetic, innovative, inquiring and exciting man, he helped produce a society displaying the very same characteristics. In the final decades of the seventeenth century, Hooke, the Royal Society, Gresham College (where the society met and Hooke lived), and the Castle Tavern (where fellows socialized after meetings) were inextricably linked in the minds of all those who understood the London scene. London would miss the sight of the diminutive genius, rushing here and there, offering an opinion on everything and trying to understand the complexities of the world.

Prejudice, Pride, and Power

Newton's propensity for protracted argument and revenge against scientific adversaries was well known. He had hounded Robert Hooke until his death because Hooke had dared to criticize publicly his early work on the nature of light. Hooke had claimed that it drew heavily on his own. In response to this accusation Newton replied:

> If I have seen further it is by standing on the shoulders of giants.

The usual interpretation of this statement is that Newton was acknowledging the place of Hooke alongside other great scientists who had inspired his thinking on optics. A less kind view is that Newton was belittling the diminutive Hooke, who was described by a contemporary as a "lean, bent, and ugly man," by excluding him from the physical "giants" Newton was prepared to acknowledge.

Such was Newton's hatred of Hooke that within weeks of taking over the presidency, Newton started making plans to move the society away from Hooke's beloved Gresham College to new premises. (The move would eventually take place, but it was a lengthier endeavor than Newton had envisaged.) He wanted to remove any obvious association of the society with the memory of his late enemy. In this he succeeded. Hooke, one of the great polymaths of a creative age, was forgotten after his death. And this was in large measure due to Newton's deliberate actions to expunge his works from the consciousness of the Royal Society and thereby from the world of science. Even Hooke's likeness was removed. When the Royal Society was based at Gresham College and prior to Newton's presidency, a splendid portrait of Hooke had adorned the wall of the society's meeting room. But this portrait "mysteriously" disappeared in the move of the society to its new premises, and no portrait of Hooke has survived to the present day. Magnificent instruments Hooke had made for the Royal Society—a microscope, lamp, air pump, and arithmetic engine—were also removed from the society. It seems likely that the actions taken to deny Robert Hooke his place in the history of science were the work of Isaac Newton. Genius, it seems, can come with a cruel temper.

Nevertheless, Newton's abilities were considerable, even as an administrator. When he assumed the presidency, finances were in disarray and meetings were poorly organized. Newton transformed the society's fortunes with his considerable organizational skills. However, once he had established his new power base as the president, he moved quickly against Flamsteed. Following the 1695 debacle over the theory of the Moon's motion, the two men had interacted again briefly on the theory in 1697, but they were unable to resolve their differences. Newton was still haunted by his failure to have solved the theory of the Moon's motion to his own satisfaction, including producing a full physical insight explainable in terms of his theory of gravitation. He had made some progress with David Gregory, in his 1700 document, but he remained convinced that Flamsteed still held the observations he needed to finally perfect the theory. He still wished to use the Moon to demonstrate the full power of his theory of gravity and to provide a signpost to the ultimate solution of the longitude problem. Moreover he saw the theory of the Moon's motion as being a centerpiece of a new edition of the *Principia* he had been planning for some time (and which Roger Cotes would edit). Thus, on a frosty morning in April 1704, he made his journey to Greenwich to visit John Flamsteed and reopen an old feud.

A few weeks after this meeting with Flamsteed, Newton arranged an audience with Queen Anne's consort, Prince George, to secure his agreement to finance the publication of Flamsteed's work. Before approaching Prince George, he assembled several of the most distinguished members of the Royal Society to be "referees" (that is, guarantors of quality) for Flamsteed's catalogue. They were indeed a distinguished group and included Christopher Wren, David Gregory, and Dr. John Arbuthnot (one of Queen Anne's physicians).

Strangely, John Flamsteed himself was not invited to join the delegation to the prince to promote the publication of his star catalogue. If Newton's intentions had been honorable, then surely the person whose labors had produced the observations that the referees promoted as "the fullest and most complete" would have been included in the visitation. It was not to be. Flamsteed must have been sorely troubled by his exclusion from the delegation, and he must have already recognized that control of his life's work could be passing from his hands.

Having been alerted by Halifax to the opportunity to curry favor with the famous Isaac Newton and the Royal Society, it is not surprising that the prince (who would not have realized how he was being used) did not need too much persuasion to fund the publication of Flamsteed's catalogue. Newton was not slow in showing the society's gratitude for this royal favor. Prince George was elected speedily to fellowship of the society. The prince in return provided Newton with the authority he needed to exert the necessary controls over Flamsteed, his secretary writing that the President was desired to take what Care in this Matter he shall think Necessary Towards the most Speedy publication of so useful a Work.

Newton now had the royal authority he needed and control over the funds that the prince would make available. His scheming was working out nicely now, and he was convinced that this time he really did have total control over the uncooperative Astronomer Royal—whom he still believed had deliberately withheld information from him in his earlier attempts to solve the theory of the Moon's motion.

There was no obvious route out of the dilemma for Flamsteed. The publication of the catalogue of the heavens could not be delayed any longer, especially as the prince was urging speedy publication. However, he was determined to ensure that the control of its contents and of its quality of production would remain with him. He had prepared a most detailed plan for the publication of all of his observations and the catalogue of calculated star positions, which would record thirty years of his achievements at Greenwich. The work was to be called, grandly, the *Historia Coelestis Britannica (British History of the Heavens)*. But Newton did not want all the observations of stars; he wanted only the catalogue of star positions and observations of comets, the planets, and the Moon to demonstrate further the validity of his theory of gravitation. He was not interested in publishing the details of the observations themselves, from which the positions of the stars had been calculated. By contrast Flamsteed felt that all his observations should be published to verify the validity and accuracy of the catalogue of star positions and to demonstrate fully the true extent of his labors. This was how Flamsteed wished his life's work to be presented for posterity.

Perfect Plans—Imperfect Planning

Flamsteed's plan saw the *Historia Coelestis Britannica* as being in three volumes. Volume I would contain a detailed preface on the Greenwich observing methods and their accuracy, plus a full justification of Flamsteed's approach. This would be the chance for Flamsteed to establish his place in the history of astronomy; modesty would not be called for. Volume I would also contain the early Greenwich observations up until 1690, obtained with the sextant originally provided with funding from Jonas Moore. Volume II would be the most precise observations Flamsteed had produced, using a telescope called a mural arc purchased with a bequest on his father's death. Here would be observations of a precision never before achieved. In Volume III Flamsteed wanted to present the actual catalogue, giving the positions of three thousand stars with accuracy dramatically better than had been reported in any previous catalogue. Since the catalogue of Tycho Brahe had contained just three hundred stars, Flamsteed's catalogue would represent a major advance in human understanding.

The husband of Flamsteed's niece, a man named James Hodgson, was asked by Flamsteed to solicit surreptitiously views on his plan for the published work from some carefully selected fellows at a forthcoming meeting of the Royal Society. Hodgson had worked for Flamsteed as an assistant, so he understood the work of the observatory and the importance of its results. What Flamsteed hoped was that if he could say that his publication plan had the strong support of some distinguished fellows, then Newton would not wish to be seen to go against their wishes. On the day of the meeting things did not go quite as intended, however. Hodgson had solicited supportive comments from a few of the selected fellows as they mingled before the meeting, and he asked for the plan to be passed to another who was beyond his immediate reach. His request was misunderstood, and the plan was inadvertently passed to the secretary of the meeting, who then read it out to the assembled fellows. It is not difficult to imagine how this could have happened in a crowded room filled with a general hubbub of conversation. Although the response of the full meeting was generally positive to the broad outline of the plan, Newton got agreement that he and the referees should have the final say on the details of the publication. Flamsteed's intention for Hodgson to gain the support of a sufficient number of selected fellows to head off any preemptive strike by Newton had failed. Newton remained in firm control of events, by chance of a simple misunderstanding of a whispered request.

Flamsteed had hoped that his plan for the *Historia Coelestis Britannica* would impress all those who read it and that they could be convinced that his accomplishments were indeed impressive. Newton and his group of referees, however, largely ignored the carefully prepared plan. Flamsteed had also hoped to include in Volume III all previous great catalogues, from Ptolemy's in the second century to Hevelius's in the seventeenth, including of course the work of his beloved Tycho, to place the Greenwich observations in their historical context and provide a complete reference work. Newton would have none of it. His priority was to get his hands rapidly on the lunar, planetary, and comet data so that he could perfect the theory of the Moon's motion and provide further verification for his law of gravity.

Flamsteed requested funds from Newton and the referees so that he could hire assistants to help him with the calculations of the star positions from his most recent observations. Flamsteed saw his star catalogue as a most valuable contribution to astronomy, bettering the combined value of all previous star catalogues by a considerable margin. His mural arc observations, however, were not yet complete; some observations had to be repeated and for others the raw data had to be converted to accurate star positions. Hence the request for wages for calculators.

Newton refused this request, but he was prepared to recommend the payment of £180 solely for the purpose of calculating "the places of the moon and planets and comets." By inference, with no money for the completion of the star catalogue, the publication would have to proceed with the catalogue in its present incomplete and imperfect form. Newton would thus be able to get his hands on the observations he needed and prevent Flamsteed from receiving the full credit his scholarship deserved for a complete and accurate catalogue.

Once the offer had been made to pay for calculators for the lunar and planetary data, Flamsteed immediately hired two such helpers. However, instead of setting them to work on the lunar observations, as Newton had insisted, Flamsteed had them calculate the star positions that he needed to help complete the catalogue. Flamsteed had accumulated over the previous thirty years some thirty thousand observations that had to be checked. Each stellar position required about two hours of careful effort to calculate. When Newton heard that Flamsteed had set his helpers to work on the stellar positions, he was outraged that his instructions for them to work only on the Moon, planets, and comets had been ignored. He refused to sanction the payment of the promised £180 (despite the fact that Flamsteed had already paid £173 to the assistants performing the calculations).

The differences in the priorities of Flamsteed and Newton are readily apparent. Flamsteed was convinced that the most important requirement for solving the longitude problem (which was, after all, the reason the Greenwich observatory had been constructed in the first place) was the catalogue of the fixed stars. He saw his star catalogue as a prized gift to scholars everywhere. Why worry about studying the motion of the "hand" of the celestial clock (that is, the Moon) if one had not first determined the precise position of the "numbers" on its dial (that is, the fixed stars)? By contrast, Newton felt that some minor uncertainty in the positions of the "numbers" could be tolerated. Any star catalogue out of Greenwich would represent a major improvement over existing catalogues that contained just a few hundred stars, so whatever the current state of the Greenwich observations the celestial "clock" dial would be much improved. What was needed, Newton felt, were the observations describing the complex motions of the "hand" across the dial of the celestial clock, so that he could use his theory of gravitation to bring a physical explanation to these complex motions. Observations of the planets and comets were needed to provide further validation of the theory of gravitation for the new edition of the *Principia*. Flamsteed's and Newton's approaches each had its own logic, and in the end both the accurate star catalogue and the precision measurements of the Moon's and planets' motions were needed. But the debate was now being driven by dogma and hatred rather than by logic.

Although Newton's initial needs were scientific, increasingly his motivation appears to have been revenge on

Flamsteed for what he believed was a deliberate withholding of information needed to develop the theory of the Moon's motion a decade earlier. In response to the oft-made challenge as to why the job was taking so long, Flamsteed told Wren, "They may as well ask why St. Paul's is not finished." (Wren's masterpiece, started in 1670, was not completed until 1710.)

It is important to remember, too, the circumstances under which John Flamsteed was working. Astronomy, of course, is a nighttime occupation. Thus from dusk to dawn, other than on the Sabbath or when the weather intervened, Flamsteed would man his telescopes. The Octagon Room, or the shed in which the mural arc was mounted, had to be opened to the night sky, so in winter he would often be working in subzero conditions. He would then have to grab what sleep he could during daylight hours. But he also had daytime duties, including tutoring boys from Christ's Hospital (and elsewhere) and overseeing the parish of Burstow. On top of his astronomical observing, his teaching, and his clerical duties, he now had the writing of the *Historia Coelestis Britannica* to deal with. Through all of this he maintained an active correspondence with fellow astronomers around the country, including Stephen Gray. Moreover, Flamsteed was never in robust health, and the pressure of his various commitments must have been great for a man of his advancing years. If he had enjoyed the active encouragement of Newton and had been promised some financial reward, then the burdens he was carrying would have been more tolerable. Instead, he was faced with Newton's hostility at every turn, and it is not difficult to understand why he felt so bitter.

Gray's letters to Flamsteed were always robust in their admiration and defense of him:

> I hope you will continue with us yet many years, and not only live to see your works published but likewise to reap some benefit of your indefatigable labours, besides that eternal commendation that will be always due to your memory. For notwithstanding the opposition and aspersions you have met with from ignorant, malicious, or unjust detractors, who may have endeavoured to depreciate your performances, posterity will know how to value them, and see that though Copernicus, Tycho, Kepler, and Hevelius have done much toward the advancement of astronomy, yet you have done much more, and not only brought to perfection but likewise made it much more immediately applicable to the most useful art of navigation.

It is likely that Gray was including Isaac Newton and his acolytes (such as Edmond Halley and David Gregory) in his sweeping condemnation of "ignorant, malicious, or unjust detractors," since they were certainly Flamsteed's principal critics. Flamsteed would have valued the comparison Gray made with his much-loved Tycho, and he must have greatly appreciated Gray's kind and supportive words at a time when his interactions with the Newton clique were becoming so unpleasant.

Negotiations on how to proceed with the catalogue staggered through 1705 and into 1706. Much to Flamsteed's chagrin, Newton and his puppet referees were quite insistent that the star catalogue in its present incomplete form must be published. Their plan was that Flamsteed's work would be printed in a single volume, but it would contain two "books." The first book would be a description of the work, the sextant observations up until 1690, and the star catalogue. The second book would be the mural arc observations up until 1705 and would include the lunar, planetary, and comet data. By insisting that everything should be in a single volume, Newton was making sure that Flamsteed could not delay the release of any material that could appear in later separate volumes in his own plan. Flamsteed knew that he could do justice to his observations only if he had time to complete them and to perfect the calculations so as to produce the definitive catalogue. But Newton insisted that the material must be handed over immediately in its present form to meet with the terms of the prince's instructions to Newton to facilitate "the most Speedy publication of so useful a Work." As a delaying tactic, Flamsteed did hand over the introductory text for the first book plus the pre-1690 observations, with which he was satisfied. But he refused to give Newton the mural arc observations and the catalogue of star positions in their existing form, since he knew them to be incomplete and imperfect. Newton would not initiate any printing until he had the catalogue in his hands. The royal command was quite clear, he insisted.

What was Flamsteed to do? He needed more time to bring the catalogue to perfection. He could see no possible reason why the publisher could not start printing the manuscript material for the first book he had freely handed over, allowing him time to finish what was left to do. Despite his dour nature and severe manner, Flamsteed was an honorable man, and he therefore agreed with Newton on what he believed would be an honorable solution. He would deposit with the Royal Society a sealed copy of the catalogue in its current incomplete form, as a demonstration of his good intent. The publisher could then start printing the Book I manuscript and early observations Flamsteed had willingly provided. The printing of this material could then give Flamsteed extra time to try to complete the observations and catalogue and remove any remaining errors—or so Flamsteed somewhat optimistically hoped. Flamsteed could then deliver the completed catalogue at some yet to be agreed on time in the future and take back the sealed incomplete catalogue. This seemed to be an honorable way out of the dilemma. But Newton did not always behave honorably. He took delivery of the sealed catalogue in March 1706.

The preparation of the manuscript for Book I finally got under way. The publisher was Awnsham Churchill, who was to be paid one pound fourteen shillings per page. Flamsteed, on the other hand, was to receive not a penny for his labors in preparing the text and in proofreading. Naturally, he was not happy with this arrangement. Newton's stinginess in refusing to sanction any payment to the author, while agreeing to a profitable involvement for the publisher, is just another indication that his motivation was revenge. Newton argued that Flamsteed's stipend from the state of £100 per annum was already payment enough, although Flamsteed's income was modest indeed compared with the £1,500 Newton typically made each year at the mint. Flamsteed bemoaned:

> It is very hard, it is extremely unjust that all imaginable care should be taken to secure a certain profit to a bookseller and his partners, out of my pains, and none taken to secure me the reimbursement of my large expenses in carrying my work above 30 years.

It had been agreed that Flamsteed would correct each page as it was typeset, and Churchill promised to set five pages a week. However, a trickle of just a single page a week slowed production to a crawl. Flamsteed would correct the proofs from Churchill overnight, then wait for the next to appear. Of course the slow printing of Book I suited his purposes, and it is unlikely that he was genuinely frustrated. It seems that during this phase, David Gregory was the most active of the referees, advising Newton on progress and quality. Since Gregory's views about Flamsteed had been clearly expressed at the time of their earlier encounter on the theory of the Moon's motion, he is unlikely to have been the most helpful or sympathetic of referees.

It was early 1708 before Churchill finished typesetting the manuscript for Book I. Now the issue of the sealed catalogue had to be faced, since Flamsteed had not yet completed the catalogue he wished to have published. On March 20, 1708, Flamsteed met with Newton and the referees at the Castle Tavern. Flamsteed argued again for delaying the catalogue to a later separate volume, as in his original plan. Newton and the referees would not agree to this proposal. Reluctantly, Flamsteed agreed to give them the mural arc observations in their current form and to correct the remaining deficiencies in the sealed catalogue provided two years earlier. In return, Newton would pay Flamsteed £125 immediately, and he would then pay the balance of the £180 promised earlier once the catalogue was rapidly put into a publishable form "as far as it can be completed at this time." In fact Newton was being miserly with the prince's funding, for when the final accounts for the project were produced they showed that he had not used all the money the prince had made available.

Fate and Treachery

For once, fate now intervened against Newton. In 1708 Prince George and David Gregory both died. Newton had lost both his royal supporter and his most active referee.

Flamsteed believed that the prince's death freed him from the royal obligation, and he again decided to play for time so that his catalogue could be completed to his satisfaction. He would not now be hurried by Newton, who could no longer hold over him the prince's plea to achieve "the most Speedy publication of so useful a Work." As far as Flamsteed was concerned, the referees had lost their purpose and directive with the prince's passing and he would no longer cooperate with them. In retaliation, Newton had Flamsteed removed immediately from fellowship of the Royal Society. (The reason given was the late payment of membership dues, although the society conveniently overlooked the more significant arrears of many other fellows; Newton himself had been seriously in arrears while at Cambridge but retained his fellowship.) Since Flamsteed had become so disenchanted with the society under Newton's presidency, loss of fellowship probably caused him little additional heartache.

Rather than worry about his fellowship, Flamsteed got on with his observations—and his correspondence with Stephen Gray, who was now back in Canterbury following his frustrating time working with Roger Cotes at the new Cambridge observatory. Newton, however, was not yet finished with his scheming. If Prince George's royal command had been invalidated by his death, then a new royal command was required. After a respectable allowance for a period of mourning, who better eventually to achieve such a command from the queen than her physician, Dr. John

Arbuthnot? Actually, the task was not so easy as Newton had hoped, but he encouraged Arbuthnot to persist. Newton still aspired to place the lunar theory, based on definitive Greenwich data, as the centerpiece of his new *Principia*.

Ultimately, Newton prevailed. At a specially convened meeting of the Royal Society council on December 14, 1710, Dr. John Arbuthnot presented a royal warrant from Queen Anne. The warrant appointed the president of the Royal Society plus any others the society wished to include as "constant visitors" to the Royal Observatory. In this context, "constant visitors" were permanent supervisors of the observatory. Having enjoyed thirty-five years of relative independence and freedom to pursue his research as he saw fit, Flamsteed would now, by royal warrant, be controlled legally by Newton and his cohorts. It seemed he was no longer to be trusted to pursue his life's calling without outside interference. Flamsteed identified his plight with that of the "noble Dane" Tycho, who had been hounded out of Denmark and had moved to Prague. "I have been treated worse than ever the noble Tycho was used in Denmark," he wrote. The visitors included Arbuthnot, another physician, Dr. Meads (who was Newton's physician), Wren, and Halley. And while Flamsteed had no reason to challenge the fairness of Wren, Arbuthnot, and Meads, adding Halley to the list left him convinced again of Newton's ill intent.

On March 14, 1711, Arbuthnot wrote to Flamsteed stating that the queen had commanded him to complete the publication of his catalogue, and that the printing must be restarted. He asked also that Flamsteed forward him immediately an updated and corrected catalogue. Flamsteed had no difficulty seeing Newton's hand behind the new turn of events and was to record:

> I was afresh disturbed by another piece of Sir Isaac Newton's ingenuity.

He wrote immediately to the secretary of state, Henry St. John, complaining about Newton's interference in the work of the observatory. But the official reply was uncompromising:

> The Queen would be obeyed.

A further letter followed from Arbuthnot, insisting that publication of the catalogue must proceed apace:

> I am the more fully persuaded you will comply with so reasonable a request, because of the regard you have for the memory of the Prince, as well as for your own reputation, both of which are interested somewhat in this performance.

Nevertheless, Flamsteed was prepared to play for time, and while showing an apparent willingness to meet the queen's wishes, he wrote to Arbuthnot suggesting that he be allowed time to work up some important recent observations and fresh discoveries to make the work even more useful. However, the reply came not from Arbuthnot but from Newton. The puppeteer had again revealed himself and was at his tyrannical height:

> By discoursing with Dr Arbuthnot about your book of observations which is in the Press, I understand that he has wrote to you by her Majesty's order for such observations as are requisite to complete the catalogue of the fixed stars, and you have given an indirect and dilatory answer. You know that the Prince had appointed five gentlemen to examine what was fit to be printed at his Highness's expense, and to take care that the same should be printed. Their order was only to print what they judged proper for the Prince's honour and you undertook under your hand and seal to supply them therewith, and thereupon your observations were put into the press. The observatory was founded to the intent that a complete catalogue of the fixed stars should be composed by observations to be made at Greenwich and the duty of your place is to furnish the observations. But you have delivered an imperfect catalogue without so much as sending the observations of the stars that are wanting, and I hear that the Press now stops for want of them. You are therefore desired either to send the rest of your catalogue to Dr Arbuthnot, or at least to send him the observations which are wanting to complete it, that

the press may proceed. And if instead thereof you propose anything else or make any excuses or unnecessary delays it will be taken for an indirect refusal to comply with her Majesty's order. Your speedy and direct answer and compliance is expected.

Newton's letter clearly hints at treason ("refusal to comply with her Majesty's orders"), a charge that would have cut to the soul of the patriotic Astronomer Royal. Worse, it was now apparent that Newton had opened the sealed, incomplete catalogue, provided by Flamsteed as a sign of honest intention, and was proceeding with publishing it.

Flamsteed could see that the fate of his life's work was now beyond his control. Newton's command of the situation seemed to be absolute. What added to Flamsteed's frustration was the realization that Newton had no intention of including the most recent mural arc observations, which represented the culmination of the Greenwich work. Flamsteed managed to get hold of a copy of one of the newly printed pages and found to his horror that his original data had been changed by Edmond Halley, who was now working for Newton on the publication of the sealed catalogue. Later Flamsteed would learn that Halley was to be paid £150 for editorial work, which rubbed salt into an already gaping wound, since he, Flamsteed, would be receiving no payment for all his extra work on the volume. Arbuthnot (having earlier lied to Flamsteed, saying that typesetting had not recommenced) tried to reassure Flamsteed that any changes Halley might make to his observations in the printed work would be made only to improve them and to please Flamsteed.

Flamsteed recognized that he could not stop the printing of the sealed catalogue, since Newton had it. But despite the royal command, he saw no reason to assist with the premature and incomplete publication of his life's work, especially with Halley playing an editorial role. Cooperation would be tantamount to assisting in the destruction of his classic catalogue. There was no point appealing to the common sense or generosity of Newton, so Flamsteed wrote instead to Arbuthnot:

> I have now spent 35 years in composing and work of my catalogue, which may in time be published for the use of her Majesty's subjects and ingenious men all the world over. I have endured long and painful distempers by my night watches and day labours. I have spent a large sum of money above my appointment, out of my own estate, to complete my catalogue and finish my astronomical works under my hands. Do not tease me with banter by telling me yet these alterations are made to please me when you are sensible nothing can be more displeasing nor injurious than to be told so.
>
> Make my case your own, and tell me ingeniously and sincerely were you in my circumstances, and had been at all my labour, charge and trouble, would you like to have

your labours surreptitiously forced out of your hands, conveyed into the hands of your declared profligate enemies, printed without your consent, and spoiled as mine are in your impression? Would you suffer your enemies to make themselves judges, of what they really understand not? Would you not withdraw your copy out of their hands, trust no more in theirs, and publish your own works rather at your own expense, than see them spoiled and yourself laughed at for suffering it?

Flamsteed revealed his intentions:

I shall print it alone, at my own charge, on better paper and with fairer types, than those your present printer uses; for I cannot bear to see my own labours thus spoiled.

So that was that. Royal command or not, Flamsteed would not cooperate in the destruction of his life's work. The president of the Royal Society could go hang; he would get no further cooperation from the Astronomer Royal, whatever the threats, whatever the bribes, whatever the scheming, and whatever the appeals to royal connections. There was no law against Flamsteed's publishing his own work at his own expense, and any publication that Newton and Halley would produce would in time be revealed as fraudulent.

Newton was not going to take this challenge to his authority lightly. Flamsteed might refuse to cooperate in the publication of the catalogue, but he could not ignore the royal warrant that established Newton as "constant visitor" to the Royal Observatory. On October 26, 1711, Flamsteed was summoned to a meeting at the Royal Society on the pretext that the visitors wished to check on the current state of the instruments at Greenwich. (Had this been the real purpose of the meeting, then a visit to Greenwich would have been more appropriate. It is more likely that Newton wanted to press Flamsteed further on the catalogue.) By chance Flamsteed encountered Halley at the entrance, who invited him to share coffee following the meeting. Flamsteed declined, with no pretense at cordiality and showed no civility in describing his disgust at Halley's role in the sorry affair. Flamsteed condemned Halley as "a lazy and malicious thief."

Flamsteed was to recount the meeting with Newton later in a letter to his friend Abraham Sharp. Flamsteed worked Newton into a rage by insisting that since the instruments at Greenwich were either gifts or had been bought with his own money, the "visitors" had absolutely no authority over them. Inevitably the discussion moved onto the subject of the catalogue, with Flamsteed accusing Newton of robbing him of the fruits of his labors:

> At this he fired and called me all the ill names, puppy etc., that he could think of. All I returned was I put him in mind of his passion, desired him to govern it, and keep his temper. This made him rage worse, and he told

me how much I had received from the government in 36 years I had served. I asked what he had done for the £500 per annum that he had received ever since he settled in London. This made him calmer, but finding him going to burst out again I only told him: my catalogue half finished was delivered into his hands on his own request sealed up. He could not deny it but said Dr Arbuthnot had procured the Queen's order for opening it. This I am persuaded was false, or it was got after it was opened. I said nothing to him in return, but with little more spirit than I had hitherto showed told them that God (who was seldom spoke of with due reverence in that meeting) had hitherto prospered all my labours and I doubted not would do so to a happy conclusion, took my leave and left them.

The path was now set. Newton and Halley would proceed with the publication of the pirated catalogue against Flamsteed's wishes and without his cooperation.

In 1712 the pirated volume, the result of the guile of Newton and the devious editorship of Halley, was published as the *Historia Coelestis*. In the preface to the work, Halley was critical of Flamsteed, stating that there were many imperfections in the catalogue he had submitted and that Halley had had to correct many of the calculations. Unkindly he criticized the time Flamsteed had taken and the expense involved, and he suggested that Flamsteed had been selfish in his endeavors. These words may well have been Newton's rather than Halley's:

> Flamsteed had now enjoyed the title of Astronomer Royal for nearly 30 years but nothing had yet emerged from the observatory to justify all the equipment and expense so that he seemed, so far, only to have worked for himself, or at any rate for a few of his friends, even if it was generally accepted that the Greenwich papers had grown into no small a pile.

The true situation regarding the sealed catalogue was misrepresented:

> The work went well and was on the point of seeing the light of day at last when the press had to stop as the catalogue of fixed stars was defective and lacked many constellations, as it had been handed over to the delegates with numerous imperfections.

Halley then promoted his own efforts while making gratuitous reference to Flamsteed's failing eyesight:

> As Flamsteed kept his eyes, now less acute at his advanced age, intent on the ever-increasing phenomena of the stars, the task was given to Edmond Halley LL.D, Savilian Professor of Geometry and thoroughly experienced in

astronomy, of supplying what the rest of the edition lacked and seeing it through to completion.

Halley, referring to himself in the third person, made extravagant reference to his own contributions:

> Not infrequently he had to correct and amend errors made through the fault of the writer [Flamsteed] or the computer [the person making the calculations] and he had to fill in quite a few gaps.
>
> Hence arose a further labour to extract the observations of individual planets, to assign them to their particular class, and in addition, to deduce the right ascensions and declinations [their position in the heavens] from them.

The observations made available to Newton in the catalogue did allow him, finally, to develop an improved lunar theory to be included in the second edition of the *Principia*, but it was merely a further refinement of the Horroxian approach. To this extent Newton's treachery achieved its selfish ends. It has been suggested, however, that had Flamsteed been left to his own devices, his definitive catalogue would probably have been completed anyway by 1712. Newton would later present to the Royal Society a leather-bound and gold-embossed version of the pirated *Historia Coelestis*, a trophy of his perceived victory over the Astronomer Royal.

But political circumstances were changing once again. In 1714 Queen Anne died, and the new monarch, George I, brought in a new set of royal attitudes. The Hanoverian monarch could not even speak English when he took the throne. The old government fell, and Newton had few contacts in the new regime. Then, in 1715, Halifax, Newton's most effective political ally, died. And while Newton had lost many of his powerful political contacts, Flamsteed had gained one, the new Lord Chamberlain, the Duke of Bolton. He recounted to Bolton the injustices he had suffered at Newton's hands, and convinced Bolton that the pirated *Historia Coelestis* was so flawed through incompleteness and errors that its survival was an affront to truth. Accordingly, Bolton signed a warrant addressed to Churchill (the printer), Newton, and the referees ordering that the 300 remaining copies of the catalogue (400 had initially been printed) must be handed over to Flamsteed. These were finally delivered to him on March 28, 1716.

Flamsteed removed the pre-1690 observations from each volume; these observations he felt were still valid and could be retained for his own publication. He then piled what remained of the 300 pirated volumes, the incomplete mural arc observations and catalogue, on a high mound in Greenwich Park and put a torch to them as his "sacrifice to Heavenly truth"

> that none might remain to show the ingratitude of two of his countrymen.

Writing to Abraham Sharp, Flamsteed noted that by taking an erroneous catalogue out of circulation he had done Newton and Halley a favor:

> If Sir I. N. would be sensible of it, I have done both him and Dr Halley a very great kindness.

Of course, various copies of the pirated catalogue already had been circulated to observatories around Europe and to libraries in England. Flamsteed wrote to recipients, pleading for their withdrawal. Even after his death, his widow was writing to those who still retained copies begging for their return. Very few of them remained in circulation.

Despite Flamsteed's moment of triumph with his sacrifice to heavenly truth, Isaac Newton still enjoyed the power of the presidency of the Royal Society; he would continue to use this power shamelessly. However, the end game for Newton and Flamsteed was nigh.

CHAPTER 5

The End Game

Sly Newton had still more to do, and was ready at coining new excuses and pretences to cover his disingenuous and malicious practices. I had none but very honest and honourable designs in my mind: I met his cunning forecasts with sincere and honest answers, and thereby frustrated not a few of his malicious designs.

—JOHN FLAMSTEED

And what was Isaac Newton up to as he entered his second decade of presidency of the Royal Society? His influence was as strong as ever. Throughout the nation his supporters still held key posts. However, his power base within the Royal Society was a little less secure than previously. Each time he was required to subject himself to reelection to the presidency, his bedrock of support was eroded further.

Many fellows were tiring of his dictatorial manner. His loathing of Flamsteed was undiminished.

Use and Misuse of Power

The second edition of the *Principia*, edited by Cotes, was published in 1713. Newton had gone through and removed almost all references to Flamsteed (and to Hooke, for whose memory his hatred remained intense). Fifteen specific references to Flamsteed were expunged. Since so many of the propositions in the *Principia* were based on Flamsteed's work, Newton had to cover himself by referring to observations from the Royal Observatory at Greenwich (without stating that Flamsteed had made the observations). Flamsteed's name was mentioned against observations of the comets of 1682 (which famously would bear Halley's name) and a comet of 1680—but Newton rather petulantly noted that Flamsteed's calculations had to be corrected by Halley:

> The following table shows the motion thereof, as observed by Flamsteed, and calculated afterwards by him from his observations, and corrected by Dr Halley from the same observations.

An improved version of the lunar theory did appear in the new edition. The second edition of the *Principia* was written in Latin, as had been the first edition. It was not until the third Latin version, published in 1726, that an English translation was produced. An unfortunate error appeared in the description of the theory of the Moon's motion in the translation. A whole paragraph was omitted in its printing, rendering its application unintelligible to even the most careful of readers. Rough publishing justice had caught up with Isaac Newton's attempt to provide the world with an ultimate theory of the Moon's motion. In his later years Newton told colleagues that he was planning a further attempt to bring a physical understanding to the theory of the Moon's motion—but he never did renew his quest.

After Newton's death mathematicians in Europe, such as Alexis Clairaut, Jean Le Rond d'Alembert, and Leonhard Euler, eventually improved the theory of the Moon's motion. They were helped by a wealth of new observations assembled by Halley and others. The first lunar tables with a precision consistently better than 1 minute of arc (and therefore providing longitudes to better than 1 degree) were produced by the German mathematician Tobias Mayer in 1753. Then the Moon could, at last, be used for determining longitude at sea with some certainty. (In fact, a £20,000 prize offered by Queen Anne in 1714 for determining longitude most precisely was not won by an astronomer. It was won by carpenter and watchmaker John Harrison, with his accurate and reliable chronometer. Harrison had his own special battle with the scientific establishment. Following royal intervention, Harrison was eventually victorious and was awarded Queen Anne's longitude prize in 1773. The French astronomer Nicolas Louis de Lacaille first used the lunar method successfully for the determination of longitude at sea in 1753. The precision achieved by Lacaille persuaded the British Admiralty to adopt the lunar method. The lunar method's heyday was probably from about 1780 to 1860. It remained attractive because it was cheaper than the chronometer method; accurate handmade chronometers were exceedingly expensive for poor seamen. In addition, the lunar method provided an important backup to the chronometer method in case of mechanical failure or accident. The *British Nautical Almanac* was published from 1767, giving lunar longitude positions at three-hour intervals. John Harrison may have secured the longitude prize, but the developments of Newton's theory of the Moon's motion and Greenwich lunar data would ensure that astronomical observations would remain an important component of accurate navigation at sea until the modern era of satellite navigation.)

In Gottfried Wilhelm von Leibniz, the distinguished German mathematician and philosopher, Isaac Newton found his intellectual match. As Newton's feud with Flamsteed waned after the "sacrifice to Heavenly truth," he redirected his energies to a long-running confrontation with Leibniz over the origin of the calculus. It was in the plague year 1666 that Newton devised what he called the "fluxional method." He generalized the mathematical methods that were being used to draw tangents to curves and to calculate the area swept by curves. He recognized that the two procedures were inverse operations. The methods became known as "differentiation" and "integration" and formed the calculus—an entirely new and powerful branch of mathematics. Newton wrote down his method of fluxions and taught it to his students on returning to Cambridge. However, he did not publish his method until much later. Leibniz independently invented the calculus in 1675 and published his methods in 1684. Newton believed that Leibniz had been alerted to the calculus by the publisher John Collins, who was aware of Newton's work, and that Leibniz had merely developed Newton's original ideas (although there is ample evidence that Leibniz arrived at the calculus totally independently). Thus what started as mild innuendos from both sides rapidly developed into unbridled hostility and claims of plagiarism. But Newton had the Royal Society to bring into play. He appointed an "impartial" committee of the society to resolve the issue. He then secretly drafted the report of the committee himself and reviewed it anonymously in the *Philosophical Transactions*. Not surprisingly, the report and its review supported Newton's claim to have invented the calculus and diminished Leibniz's role. Newton and his acolytes continued to hound Leibniz shamelessly until his death. But while Flamsteed and Leibniz continued to attract Newton's fury, Stephen Gray continued with his astronomical observations and would make an unexpected low-profile reentry to the life of the Royal Society.

Sun Worship

Despite his constant fatigue from the demands of dyeing, Stephen Gray was keen to make astronomical observations of a variety of phenomena for his friend Flamsteed. Gray's sunspot observations in the decade following his return from Cambridge are of particular interest. The Sun cannot be observed with the unaided eye without risking blindness. However, since antiquity the Sun's disc has been viewed with relative safety when reflected off water, seen through heavily smoked glass, or when its brightness is highly diminished and reddened at sunset or sunrise. With the discovery of the telescope, images of the Sun could be projected onto a card and viewed without danger to the eye.

The brilliant disc of the Sun occasionally displays dark regions, sometimes large isolated spots, sometimes clusters of several smaller spots. The frequency of sunspot occurrence waxes and wanes over an eleven-year cycle, the so-called "sunspot cycle." Sunspot activity is usually associated with the ejection of giant solar flares from the Sun's surface. The sunspot cycle is more than a mere scientific curiosity, since it is a measure of the Sun's activity and is related indirectly to the Earth's climate.

The sunspot cycle sometimes switches off, for decades or even centuries. Sunspots occur only very rarely during such a period of quiescence. These intervals of suppressed sunspot activity coincide with periods of extremely low global temperatures, popularly referred to as "mini–Ice

Ages." From about 1630 to 1705 such an interval of low sunspot frequency occurred. This seventeenth-century period of low sunspot activity was recognized from historical data by a nineteenth-century astronomer called Walter Maunder. It now carries his name—the "Maunder Minimum." Several earlier such minima have been established from historical observations of sunspots. Their linkage to climate variations can be demonstrated by certain indicators of historical climate, such as the width of growth rings in trees of extreme age and in fossilized wood. The Maunder Minimum was a period of very low winter temperatures in Europe. Unbeknownst to Stephen Gray, his sunspot observations on his return from Cambridge were tracking the emergence of the Sun from the Maunder Minimum and heralded an epoch of milder winters. He described his sightings of sunspots with enthusiastic surprise, emphasizing their rarity at this time.

Gray was fascinated by the appearance of the giant spots he detected, and he described in detail their changing morphology. He described characteristics of spots that were not described again until the modern era. The following account submitted to the Royal Society showed an imaginative interpretation of the dark spots and bright regions (called foculae). It was one of the letters whose publication was suppressed during the Newton presidency:

> Before I conclude give me leave to add a few conjectural explications of some of the phenomena of these spots

which I shall do by running a parallel betwixt the solar and terrestrial fires. The foculae do not appear, as was said, but when near the limb of the sun, and the nearer thereto the brighter, the reason thereof I take to be that property of flame to appear brighter when most contrasted. The spots being thrown up so suddenly argues that they are ejected with a most violent force, as those that are sensible to their magnitude will easily believe. They seem to proceed from the internal parts of the sun being cast out from thence in the same or some such like manner as sand stone and ashes etc are thrown from Mount Etna.

Flamsteed greatly admired Gray's ability as an observer and made extensive use of his observations of sunspots and eclipses. This is apparent in his defense of Gray when an error appeared in some solar eclipse results sent to the Royal Society by a Dr. Harrys. In writing to a colleague, Flamsteed criticizes Harrys as merely a "transferrer or translator" of the results of others. Gray had helped Harrys with the observations in which the error had been made. Flamsteed noted:

Mysteriously there is a mistake of a whole minute or more omitted in counting the clock and of the principal phases. This I am apt to believe Mr Gray will not own to be his fault, he was too cautious to commit such a one. But the Dr who owns himself not so expert must bear the blame of this not being all to rights, and set the saddle on the horse it belongs to.

The strangest example of Flamsteed demonstrating his respect for Gray was his asking him some years earlier to help investigate a ghost story. The city of Canterbury had been much exercised by the story that one of its more sober residents, Mrs. Bargrave, had met with an old friend, a lady called Mrs. Veal, who (unbeknownst to Mrs. Bargrave) had recently died. Mrs. Bargrave was adamant that she had entertained Mrs. Veal in Canterbury the day after she had died in Dover (and there seemed to have been no way that Mrs. Bargrave could have been alerted to the death of her friend). There was much talk of witchcraft. Mrs. Flamsteed had been alerted to the story in a letter from a friend in Canterbury who signed herself "E.B."

John Flamsteed shared the strange ghost story related in the letter from "E.B." with Dr. John Arbuthnot, Queen Anne's physician. Arbuthnot then wrote to Flamsteed:

Sir, I was asked the other day by a very great person if I had heard anything of the story you showed us in your letters about the apparition at Canterbury. I said I had, and mentioned the letters that you had. I also added that I believe I could procure a copy of them, which I beg you would do me the favour to send me by the penny post (direct for me at my house in St James Place), with what you know of the credit of the persons concerned. I shall not give the copies to any person, but them I mention; nor shall it be published by my allowance. In doing this you will extremely oblige. Sir, Your most humble servant. Ja Arbuthnot.

It is likely that the "very great person" was Queen Anne.

Recognizing the importance of the commission from Arbuthnot, John Flamsteed wrote to his trusted friend Stephen Gray asking him to investigate the details and the veracity of Mrs. Bargrave's story and to report his opinion on "the credit of the persons concerned." This Gray did with characteristic diligence and the urgency demanded by a royal request (despite the demands of his business). He reported his research on the story in the careful, dispassionate tone of the gifted experimental scientist. His letter giving his description of the apparition was long and thorough in its analysis. He was not jumping to conclusions, however, about the veracity of Mrs. Bargrave's story. He noted that Mr. Bargrave had told him that his wife had claimed she had sighted ghosts previously!

The story is, of course, of no great interest in itself other than to illustrate the fact that in the early eighteenth century, ghost stories attracted royal attention, and explanations in terms of witchcraft could even be contemplated. As a demonstration of the trust the Astronomer Royal was prepared to place in Stephen Gray, however, it is important. A copy of Gray's reply to Flamsteed was passed to Arbuthnot and presumably thence to the queen. The Veal-

Bargrave affair eventually found its way into the writings of Daniel Defoe (the author of *Robinson Crusoe* and *Moll Flanders*) and so is preserved in the English literature of ghost stories.

London Beckons

Although it is difficult to establish all the facts with precision, it does seem that Stephen Gray stayed in London for a period from 1715 to 1719, until he entered the Charterhouse (a home for poor pensioners) in 1719. There are no extant Gray letters written to Flamsteed from Canterbury after December 1714. One possible explanation for the lack of letters is that, living in London, Gray was able to make the easy river journey to Greenwich to visit Flamsteed, and therefore there was no need for written correspondence.

A further piece of evidence supporting the notion that Gray resided in London from 1715 is the Journal Book for Royal Society meetings, which records that "Mr. Gray had leave to be present" on several occasions from March 1, 1715, until December 15, 1720. (Non-fellows were allowed to attend meetings of the society with agreement.) Had he still been running the dyeing business in Canterbury at this time, it is unlikely that Gray could have afforded either the time or money for travel from Canterbury to attend meetings of the society.

William Stukeley suggests that Gray stayed at this time with Jean Théophile Desaguliers in a house on old

Westminster Bridge in London. Desaguliers's father had been a Huguenot pastor who had to escape persecution in France. He made his passage to England with his young son hidden in a tub. The young Desaguliers was a gifted scholar and graduated in mathematics from Oxford. He eventually became friendly with Newton, who appointed him as a demonstrator at the Royal Society. Gray is reported to have helped Desaguliers with his experiments at meetings of the society and with astronomical observations from the house on old Westminster Bridge. (The medieval bridges across the Thames not only provided river crossings, but also were crowded with houses and market stalls.) Perhaps Gray moved to join Desaguliers in London between his final letter from Canterbury in December 1714 and the first record of his attendance at the Royal Society in March 1715. From 1716 until 1719 there are frequent notes in the Royal Society Journal Book of Desaguliers reporting astronomical observations made by Gray and himself at Westminster.

It is interesting to note that while Newton had earlier suppressed Gray's research results because of his friendship with Flamsteed, Gray was allowed to help Desaguliers with demonstrations at the society. Perhaps Newton was willing to allow Gray to perform menial tasks for the society as an assistant to a demonstrator but not to allow him the accolades due from published research results or full fellowship. Indeed, having Gray, an outspoken supporter of Flamsteed, as a servant of the society could have provided a smug Newton with a degree of satisfaction. While Newton's acolytes were accepted into full fellowship in the society and into prestigious academic appointments, a Flamsteed acolyte could be seen to progress no further than mere service to the society. Also at this time no doubt Newton felt he had now won the battle with Flamsteed over publishing his results (despite Flamsteed burning so many copies of the disputed star catalogue), since he did get the observations he wanted.

There is an alternative explanation for Gray's attendance at meetings of the Royal Society during the period 1715 until 1720. Could Gray have recanted his support for Flamsteed and thrown in his lot with the Newtonians? This would explain both the end of his correspondence with Flamsteed and his now being allowed to attend meetings of the society with Newton in the chair. However, Gray's support for Flamsteed had been so solid and his behavior so honorable in all other respects (both before and after this time) that such an explanation seems highly implausible. Moreover, following Gray's final letter from Canterbury in December 1714, there are two letters to Flamsteed in January 1716 from Norton Court, where Gray was visiting Godfrey to undertake astronomical observations. The letters are friendly and end with the usual felicitations. "Mr Godfrey gives his humble service to you, as does he who is Sir your humble servant." Thus, a souring of the relationship between the two men seems unlikely; however, since there is no written evidence of rapport after the letters from Norton Court, we cannot be sure.

Gray did write a paper in 1720 that was published in the *Philosophical Transactions*. It was unlike any other submitted after the start of the Newton presidency. It reported the discovery of various nonrigid electric materials, and it was the only piece of Gray's published during the Newton presidency. It is not obvious why this single letter was selected for publication or why Gray had returned to experiments in electricity. The new electric substances Gray identified included "downy fibre of feather," hair from Gray's wig, "the fine hair of dog's ear," "silk of several colours and of several finenesses," "pieces of ribbon," "linen of several sorts," "paper both brown and white," "shavings of wood," "leather, parchment," and "ox guts wherein leaf gold is beaten." All were electrified by rubbing.

Flamsteed's Final Move

By 1716, Flamsteed's health was deteriorating rapidly, but he was intent on getting his definitive work completed for the press. He had reverted to his original publishing plan of 1705, involving three volumes. The first volume was to contain a grand introduction describing in detail the observing methods at Greenwich and their place in history. Flamsteed completed the volume in English, but it needed to be translated into Latin (traditionally used for scholarly treatises). The first volume would also contain all the pre-1690 sextant observations. Flamsteed had proofread and corrected them for the earlier version and was content that there was no need to rework them. The ninety-seven pages from the pirated Newton-Halley edition could be rebound in the new definitive volume. Volume II was to contain all the mural arc observations conducted from 1690 until the time of publication. Here would be the planetary, lunar, and comet observations coveted by Newton. Flamsteed made steady progress in calculating the star positions from the mural arc observations and in preparing the volume for the press. He had made scant progress with the third volume, the star catalogue, when his condition deteriorated further, and he died at his beloved observatory on the last day of the year 1719. He had been making observations until a few days before he died. Flamsteed had written, "God suffers not man to be idle, although he swims in the midst of delights." Like Isaac Newton, John Flamsteed had never been idle.

With a life dominated by illness, it is tribute to Flamsteed's indomitable spirit that he survived to the respectable age of seventy-three years, forty-four of them as Astronomer Royal—he had served six monarchs! Flamsteed was buried in front of the altar at the Burstow church he had served faithfully as rector for thirty-five years.

That might have been the sad end of John Flamsteed's story, had it not been for the dedication of his widow, Margaret, and his loyal assistant, Joseph Crosthwait. They vowed to complete Flamsteed's monumental undertaking. Crosthwait enlisted the assistance of Abraham Sharp, another former assistant at Greenwich, and the two men, with Crosthwait bearing the larger burden over a five-year period, completed the preparation of the volumes for the press. Flamsteed's *Historia Coelestis Britannica* was finally published in 1725, fifty years after the foundation of the Royal Observatory at Greenwich, dedicated to the task of producing the definitive catalogue of the stars. This was certainly the astronomical equivalent of the construction of Saint Paul's Cathedral, and it established Britain's place at the forefront of world astronomy and Greenwich's place in the history books. It also secured Flamsteed's reputation as a worthy successor to "the noble Dane." His forty-four years of commitment to a single task, despite all the hardships he endured, must surely stand among the greatest accomplishments of the human spirit.

The *Historia Coelestis Britannica* stands as testimony to John Flamsteed's skill as the most able astronomer of the seventeenth century (while the pirated catalogue stands as testimony to the treachery of Isaac Newton and Edmond Halley). Such was the impact of Flamsteed's great work that Greenwich was finally adopted in 1884 to be the reference point for measuring longitude. The zero meridian passes through the site of Flamsteed's studious observations.

Although Flamsteed originally drafted a preface to his work that was critical of Newton, his executors thought better than to include it in the published edition and thus cause a further confrontation with the Royal Society. Even with Flamsteed in his grave, Newton could still influence his work. The unpublished preface includes the following damning indictment of Newton's methods:

> His design was to make me come under him . . . force me to comply with his humours, and flatter him, and cry him up as Dr Gregory and Dr Halley did. He thought to work me to his ends by putting me to extraordinary charges. Those that have begun to do ill things, never blush to do worse to secure themselves. Sly Newton had still more to do, and was ready at coining new excuses and pretences to cover his disingenuous and malicious practices. I had none but very honest and honourable designs in my mind: I met his cunning forecasts with sincere and honest answers, and thereby frustrated not a few of his malicious designs. I would not court him. For, honest Sir Isaac Newton (to use his own words) would have all things in his own power, to spoil or sink them; that he might force me to second his designs and applaud him, which no honest man would do nor could do; and, God be thanked, I lay under no necessity of doing.

The task of Crosthwait and Sharp was not yet completed. They oversaw the production of the engraved plates showing the constellations with the accurately determined star positions. Flamsteed's *Atlas Coelestis*, the largest star atlas ever produced, was published in 1729. (The atlas, measuring 24 by 20 inches, was so large that it was not easy to use. In addition, some of the constellation engravings were not felt to be as artistic as in previous classic catalogues. Astronomers in France modified Flamsteed's catalogue in 1776, producing it in smaller format and with more pleasing artwork. Their *Atlas Céleste* became the standard version of Flamsteed's atlas in common circulation and was used by European mariners for the next hundred years.)

In 1603, the astronomer Johann Bayer had introduced a system of using Greek letters to name the stars in a constellation, with alpha being used for the brightest star, beta the next brightest, and so on. Halley introduced a numbering system , numbering the stars from west to east in each constellation regardless of brightness. Bayer had reversed the representations of the constellations used since Ptolemy's time, depicting them as if viewed from the outside of a sphere (such as viewing a globe of the Earth). The traditional method was as if viewing from inside a sphere—in other words, as if viewing the night sky lying on one's back. Thus, for example, Ptolemy's star on the right shoulder of Orion had become in Bayer's rendering the star on the left shoulder. Flamsteed was determined to right the Bayer representation and reintroduce the Ptolemy tradition. This Crosthwait and Sharp did for the *Atlas Coelestis*.

The preface to the *Atlas Coelestis* ends with two paragraphs that capture some of the frustrations experienced in producing the work, along with the pride of John Flamsteed's wife and friends in his achievements.

As works of this nature meet with very few encouragers, and as a great part of the Historia Coelestis, as well as this book, have been carried out at the sole expense of the executors, they were unwilling to proceed in one until the other was published, which, together with the difficulties and delays that usually attend performances of this kind, has been the reason why it has not appeared sooner abroad; but as neither pains or expense have been wanting to render it as complete as possible, there is reason to hope that it will meet with suitable reception from the generous, candid, and unprejudiced part of mankind.

And lastly, as the principal view of the Royal founder of the Observatory was to obtain a good catalogue of fixed stars, so it must be justly acknowledged that Mr Flamsteed has fully accomplished that great end, having left behind him one of the largest and most complete catalogues that ever the world was enriched with, from which these charts are deduced, containing almost double the number of stars in that of Hevelius, to the honour of the British nation, and the lasting reputation

of the author; a work that will render his name famous
to the latest posterity, and perpetuate his memory until
time shall be no more.

The history and atlas did indeed render Flamsteed's name
famous—and perpetuate his memory "until time shall be
no more"—despite the efforts of Sir Isaac Newton.

The remarkable dedication of Crosthwait and Sharp
in completing Flamsteed's work placed them in astronomy's annals of fame. Theirs was entirely a labor of love,
since neither received any payment from Mrs. Flamsteed
(although it seems Crosthwait, whose role was the greater,
had been led to believe that some payment from Flamsteed's estate would be forthcoming). It appears likely that
Mrs. Flamsteed lost much of her considerable inheritance
from her husband in the "South Sea Bubble." This was a
share scam based on the South Sea Company's trading in
slaves to Spanish South America. The company offered to
take over a portion of government debt in return for future
trading guarantees. Bribed government ministers talked up
the share price, thus attracting interest from a large number
of inexperienced small investors such as Mrs. Flamsteed.
The inevitable collapse of the overpriced stock, in September 1720, ruined many small investors. It could be that
Mrs. Flamsteed failed to pay Crosthwait not out of mean-
ness but out of an unexpected state of poverty.

Even before Flamsteed's death, plans were afoot for him
to be succeeded as Astronomer Royal by his archenemy

Edmond Halley. Other than Newton himself, no person could have caused Flamsteed more distress as his successor. Halley left no time in ordering Mrs. Flamsteed to vacate the observatory. His hurried takeover was to no avail, however, since there was no observing to be done on his arrival. Margaret Flamsteed had stripped the observatory of all its equipment; she claimed that all of it had been given to John Flamsteed as personal gifts (for example, by Jonas Moore) or had been purchased by Flamsteed with his own money. Halley's appeals for the return of the equipment were in vain, and the government eventually decided not to press the executors since their case was undoubtedly sound. Instead, Halley was made a grant of £500 by the government (equivalent to the original cost of the observatory) to reequip himself with a new and better mural arc. It was 1724 before the new instrument was ready for observations.

Gray's Next Move

From shortly after Flamsteed's death, there is no reference to Stephen Gray's attending any meetings of the Royal Society until 1730—some years after Newton's death. Perhaps the death of Flamsteed and the memory of Flamsteed's humiliation at the hands of Newton left him sufficiently depressed that he could no longer face the prospect of attendance at the society with Newton in the chair. By this time it is recorded that an aging and cantankerous Newton slept through many of the society's meetings.

Stephen Gray had first approached Hans Sloane, secretary to the Royal Society, in 1711 to intercede on his behalf to find a place at the Charterhouse, a home for poor pensioners. The name "Charterhouse" is a corruption of the French name "Chartreus," the location of the first Carthusian monastery. Houses of the Carthusian order were established in medieval London, and one of them eventually became a home for pensioners, combined with a school. It took the Charterhouse name. Its constitution stated:

> There shall no rogues or common beggars be placed in the said Hospital but such poor persons as can bring testimony and certificate of their good behaviour and soundness in religion.

Gray's letter to Sloane asking him to intercede on his behalf with the Charterhouse governors was a heartrending account of his unfortunate state of health and poverty:

> Sir, It may I fear be accounted too great a presumption in me to give you the trouble of this address. But the experience I often had of your candour towards me when I have made bold to communicate to you my mean thoughts and experiments on some philosophical subjects did encourage me to think you will pardon this bold attempt if you cannot comply with my request, especially considering I am necessitated there to and

know no fitter a person, Sir, than yourself to apply
myself to on this affair.

I have for many years spent the far greatest part of my
time that the avocations for a subsistence would permit
me in the study of astronomy, and not without making
inspections into some other parts of experimental knowledge. And my mean circumstances considered been at no
little charge for books, instruments and other materials,
being prompted thereto chiefly by the natural tendency
of my genius, though not altogether without hopes that
some greater advantage might at one time or other
attend than barely the satisfaction of my inclinations.
But I find it otherwise, and now being in my 45th year
of my age think it time to consider how I shall procure a
comfortable subsistence, being already so infirm as not
to be able to follow my employ without much more
difficulty and pain than in former years caused by a
strain I received in my back some years ago which brought
on me the Dolor Coxendicis. Now I am prompted by my
friends to believe that if application were made to some
of the Governors of the Charterhouse on my behalf, they
would grant me a warrant to be admitted there. Sir, this
comes therefore humbly to desire you would be pleased
to intercede for me to my Lord Penbrook, my Lord
Summers, or any of the Governors you are best
acquainted with. If so great a favour might be obtained,
I should think myself happy and those many and great

interruptions I now meet with would be removed so that I should have time enough to make a further progress than I have yet done in some enquiries relating to astronomy and navigation and might happily find out something that might be of use.

Sir, if you shall think fit to use your interest for me in this affair I shall be ever obliged to acknowledge the favour and shall endeavour to serve you in whatever my ability will permit.

Despite the best endeavors of Sloane and others, Gray was not nominated for a place among the eighty "gentlemen pensioners" of the Charterhouse until a meeting of the governors on June 24, 1719—when his nomination was by "ye Prince" of Wales. Perhaps, having failed to get Gray a place at the Charterhouse in 1711, Sloane persuaded Desaguliers to offer him a home at Westminster Bridge in return for undertaking menial tasks for the society.

Although the Charterhouse enabled a man to live without fear of starvation, it certainly did not provide a luxurious life. However, the standard would have compared very favorably with life in the poorhouse. The basic pension was £5 a year. Pensioners had their own small rooms surrounding a quadrangle. It must have reminded Gray of his time at Trinity College, on a very much more modest scale. A midday dinner (meat, bread, broth, and ale) and a supper were provided—but no breakfast. In 1730 the governors noted:

> They have no allowance for breakfast. It is a great hardship to fast from supper at six o'clock to dinner . . . and therefore humbly praying us to make them some allowance for breakfast, we do order that the said pensioners if they think fit may take the broth provided for them at dinner in the Hall and carry it to their chambers for breakfast the next day.

Rules abounded at the Charterhouse. Gowns were to be worn at all times, pensioners were required to attend morning prayer at 11:00 each day, and permission had to be gained to stay overnight elsewhere. This last rule did not seem to inhibit Gray from spending time with John Godfrey at Norton Court or from about 1727 with another gentleman scientist, the Reverend Granville Wheler, at Otterden Place. Presumably Flamsteed had introduced Gray to Godfrey, aware of the two men's interests in both astronomy and electricity.

After his grueling life as a dyer, the Charterhouse seems to have appealed to Gray. He enjoyed the company. A fellow pensioner wrote:

> We did eat and drink together on Saturdays and Tuesdays. . . . Mr Gray would smoke two or three

pipes and gave me a great deal of delight and satisfaction in his very agreeable conversation.

From 1720 until 1728, little is known of Gray's activities at the Charterhouse. He seems to have kept a low profile in the affairs of the institution. The order books of the Charterhouse give the minutes of the governors' meetings of the period, and many pensioners are mentioned by name. Some were troublemakers; others petitioned the governors on various matters. Gray's name does not appear in the order books. There are no letters from him to the Royal Society or to others that can reveal what avenues of science he was pursuing at this time. We must assume that he continued to conduct his astronomy and electrical experiments with vigor, although the Newton presidency still meant that wider recognition was not possible.

Newton's Final Move

By 1724, the health of the octogenarian Newton was starting to fail. His visits to the mint and the Royal Society became less frequent. Control of the mint was passed to his deputy, John Conduitt, who had married Newton's stepniece Catherine Barton after the death of Halifax. Newton moved to a new residence in Kensington, in the countryside outside London, in the hope that the fresher air would improve his health. He now had some difficulty walking, and during his increasingly infrequent visits to

London he was transported everywhere in a Bath chair. But still he refused to resign from the presidency of the Royal Society.

Newton finally succumbed to a bladder stone on March 20, 1727, after two weeks of great suffering. John Conduitt recorded his final minutes of pain:

> He rose to such a height that the bed under him, and the very room shook with his agony, to the wonder of those that were present. Such a struggle had his great soul to quit his earthly tabernacle!

Despite the pleas of John and Catherine Conduitt, Newton refused to accept the last rites of the Church of England. John Conduitt, who later wrote extensively on his step-uncle's achievements, justified this refusal on the basis that Newton's life had been so pure, flawless, and close to perfection that no preparation for paradise was necessary. Conduitt described Newton's life as

> one continued series of labour, patience, humility, temperance, meekness, humanity, beneficence and piety without any tincture of vice.

This was the start of the process of eulogizing Newton's memory and expunging from history the true record of his tyrannical traits and intellectual extremes.

Newton was buried in Westminster Abbey, an honor reserved for the nation's greatest heroes. The funeral was recorded by the *London Gazette*:

> On the 28th March 1727, the corpse of Sir Isaac
> Newton lay in state in the Jerusalem Chamber, and
> was buried from thence in Westminster Abbey, near the
> entry into the choir. The pall was supported by the Lord
> High Chancellor, the Dukes of Montrose and Roxbor-
> ough, and the Earls of Pembroke, Sussex, and Maccles-
> field, being Fellows of the Royal Society. The Hon. Sir
> Michael Newton, Knight of the Bath, was chief mourner
> and was followed by some other relations, and several
> eminent persons, intimately acquainted with the
> deceased.

Voltaire attended the ceremony and noted that it was an occasion worthy of a king. Isaac Newton had been indeed the "King of Science," but he had also unfortunately been a tyrannical dictator of scientists.

The way was now clear for Stephen Gray to emerge from Isaac Newton's shadow so that his true genius could be realized and he could finally gain respect and recognition.

CHAPTER 6

Respect and Recognition

Mr Stephen Gray has made greater variety of electrical experiments than all the philosophers of this and the last age.

—JEAN THÉOPHILE DESAGULIERS

The two men made up an unlikely partnership. Stephen Gray now looked considerably older than his sixty-two years, and he was suffering from poor health. His slight frame resulted from the dearth of dietary variation offered to pensioners of the Charterhouse. The wrinkled skin of his craggy hands testified to the years of toil as a dyer. By contrast, the younger man's hands showed no evidence of physical endeavor. They were the tender hands of a gentleman preacher, the Reverend Granville Wheler.

Electrical Experiences

The date was July 14, 1729. The location was the Great Barn at Wheler's magnificent country residence, Otterden Place, in the county of Kent. Stephen Gray and Granville Wheler were about to perform an epoch-defining experiment in electrical communication. For Gray this would be a defining moment in a lifetime of scientific endeavor. It had been a lifetime of highs and lows. Friendship and collaboration with the Astronomer Royal, John Flamsteed, represented a true high point, but unwarranted rejection by Sir Isaac Newton that resulted in the neglect, even theft, of his research by the scientific establishment was a definite low.

Gray had started his electrical communication experiments in a modest way at the Charterhouse. Almost twenty years had passed since his first tentative experiments with electricity at Cambridge and the heartbreak of their being ignored by the Royal Society.

At the Charterhouse, Gray had acquired a new cylindrical glass tube to use to generate electricity by rubbing. At 3 feet in length and 1.2 inches in diameter, it was larger than the tube he had used in Cambridge. He had corked the tube at each end to keep out dust, but these stoppers were about to lead him to a very significant discovery:

> The first experiment I made was to see if I could find any difference in its attraction, when the tube was

stopped at both ends by the corks, or when left open; but could perceive no sensible difference. But upon holding the down feather over the upper end of the tube, I found that it would go to the cork, being attracted and repelled by it as by the tube when it was excited by rubbing. I then held the feather over against the flat end of the cork, which attracted and repelled many times together, at which I was much surprised and concluded that there was certainly an attractive virtue communicated to the cork by the excited tube.

As with so many accidental discoveries in science, this one depended on Gray's recognizing that something unexpected had happened that needed to be investigated further. The significance of his chance discovery was profound. It seemed that the "stuff" of electricity (the "virtue" in eighteenth-century terminology) was not confined to the body that had been electrified by rubbing, as was previously thought, but could "flow" to a connected body. Gray was the first to demonstrate this fact. He wanted to explore the unexpected result further:

Having by me an ivory ball of about one inch three tenths diameter with a hole through it, this I fixed upon a fir stick about four inches long; thrusting the other end into the cork and upon rubbing the tube found that the ball attracted and repelled the feather.

The ball was then connected to longer rods—first one 8 inches long and then one 2 feet long. The effect was the same. Then the ivory ball was replaced with other objects—a cork ball, a kettle, fire tongs, and other simple items easily at hand. Still the electric "virtue" excited in the glass tube could be communicated to the connected object. Gray had shown that the electric effect could be produced in the glass tube (referred to as the "transmitter" of the "virtue") and then passed along a suitable communication path to another object (referred to as the "receiver"). A very modest start in electrical communication had been made. This communication seemed to be dependent on generating an electrical "signal" in a "transmitter," sending the signal through a "communication path," and detecting it at a distant "receiver." Of course Gray did not have a vision of the telegraph, the telephone, the fax, or e-mail when making these early experiments. However, he did demonstrate for the first time the fundamentals of electrical communication.

Gray now realized that the challenge was to see how long the communication path could be made. A more sensible detection method was needed than a down feather, so Gray started using brass hammered into thin leaf, placed on a length of wood. When the wood was held up to the object receiving the "virtue," then the leaf brass would be attracted toward the object in the same way as a down feather.

At the Charterhouse, Gray managed to transmit electrical effects along a communication path of just 18 feet, a distance limited by the space available to him in his chamber (although later he would get permission to set up poles in the courtyard to increase the communication path he could use). He needed somewhere with much more space to carry out his electrical communication experiments, and his wealthy friends John Godfrey and Granville Wheler could provide that. Gray had often visited Godfrey at Norton Court at Flamsteed's request for joint astronomical experiments and was confident Godfrey would help him.

The origin of Gray's friendship with Wheler is uncertain, but it might have been through Godfrey, who seemed to be well connected to the other gentlemen scientists of the time. Gray described Wheler as "a worthy member of the Royal Society with whom I have had the honour to be lately acquainted." Wheler had acquired the ideal site for communication experiments the previous year—a grand mansion called Otterden Place. The stately rooms of Otterden Place could provide spacious laboratories, and Wheler was eager to help.

Gray had started his 1729 country visits eight weeks earlier than his historic experiments of July 14. Visiting Godfrey at Norton Court on May 14, 1729, he showed that by connecting his glass cylinder to a rod 24 feet long, with a cork ball at the end of the rod, the ball would attract leaf brass when the glass cylinder was rubbed. He wrote:

> The ball attracted and repelled the leaf brass with vigour, so that it was not at all to be doubted but with a longer pole the electricity would have been carried further.

On May 16, Gray and Godfrey repeated the experiment with a pole 32 feet long, and the effect was the same. But how could the communication path now be lengthened to the hundreds of feet Gray thought might be possible? Three days later Gray replaced the solid rod connecting the glass tube and the sphere with a 26-foot length of packthread,

> which was the height I stood at in the balcony from the court where he that held the board with the leaf brass on it stood.

Again the leaf brass was attracted to the ball. The communication path was now lengthened by connecting a 34-foot length of packthread to an 18-foot pole held up from the balcony like a fishing rod. Now the total communication path was 52 feet, and still the leaf brass was attracted, although

> these experiments are difficult to make in the open air, the least wind that is stirring carrying away the leaf brass.

Gray then tried to use a horizontal communication path. He made loops at various points in the packthread, which he hung over nails in a roof beam. But now the experiment did not work, regardless of how energetically he rubbed the glass tube. Something was going wrong when packthread was used as a horizontal communication path, with the thread making direct connection to the supporting structure. Gray put the failure of the experiment down to not using proper materials and methods:

> Upon this I gave over making any further attempts at carrying the electricity horizontally, designing at my return to London if I could get assistance to have tried the experiment from the top of the Cupola of St. Paul's, not doubting but the electric attraction would be carried down perpendicular from there to the ground.

By early July Gray was back in the country, this time visiting Granville Wheler. He had taken his glass tubes and other equipment with the intention of demonstrating his simple communication experiments to Wheler. He had not expected to extend the experiments, but Wheler was keen to take on new challenges. The first experiments conducted by Gray and Wheler at Otterden Place, in the first few days of July, repeated the earlier experiments made at Norton Court. This involved transmitting electrical effects along a communication path of packthread suspended vertically from the gallery of the Great Hall (a distance of

16 feet), then from the battlements (a distance of 29 feet), and eventually from the Clock Tower (a height of 34 feet). That was the greatest vertical drop available to them. The electrical effects flowed without apparent impediment along these vertical distances. According to Gray, Wheler was fascinated by the experiments and it was he who

> was desirous to try whether we could carry the electric virtue horizontally.
>
> I then told him of the attempt I had made with that design but without success, telling him the method and materials made use of. He then proposed silk threads to support the line. I told him it might do better upon account of its smallness, so that there would be less virtue carried from the line of communication.

The Matted Gallery at Otterden Place, in which their first horizontal experiment was performed on July 2, was 80 feet long. Imagine the strange scene: the ornately paneled grand gallery, lined with magnificent paintings, crossed with fine silk threads tied to nails hammered into the picture rail circling the gallery. A communication line of packthread was suspended horizontally, lying across the horizontal lengths of silk thread. The experiment worked perfectly, and the leaf brass deflected strongly. So the limit to communication was clearly greater than 80 feet:

This experiment succeeded so well, and the length of the
Gallery not permitting us to go further in one length,
Mr Wheler thought of another experiment.

Wheler proposed doubling the line back, thus successfully transmitting electrical effects over a return distance of
some 147 feet. More space was now needed if even longer
communication paths were to be achieved. Wheler suggested
that the experiments should be moved to the Great Barn.

The Great Barn was a massive building, brick based
with wooden upper structure and tiled roof, built for hous-
ing stock and feed over the winter. Here a communication
path of 124 feet could be achieved in a straight run. Again
the communication line was laid across supporting silk
threads run across the barn to connecting nails. There
would have been less concern about hammering nails into
the walls of the barn than about hammering nails into the
fine paneling of the Matted Gallery!

On July 3 the experiments produced another unexpected breakthrough—literally. A return path was first
achieved within the Great Barn, producing the longest
communication path yet, of over 200 feet. To achieve a
greater length, a third return was attempted. However,
the silk threads supporting the communication line were
not strong enough to support a third length, and they
broke. Gray had brought some iron wire with him of a
similar thickness to the silk thread, so the two men

worked diligently to replace the silk with the thin iron wire to support the communication path. But now their experiments stopped working. Try as they might by energetic rubbing of the glass tube, they could not get a deflection of the leaf brass at the remote sphere:

> By which we were now convinced that the success we had before depended upon the lines that supported the line of communication being silk and not upon their being small, as before trial I imagined it might be.

The iron supporting wire was apparently carrying the "virtue" away to the barn walls, rather than to the distant receiving sphere. Gray had revealed for the first time the distinction between electrical "conductors" (which would easily carry the "virtue") and electrical "insulators" (which would stop the unwanted leakage of the "virtue"). Metals were found to figure prominently among the conductors, while silk thread was acting as an insulator. Here was another significant discovery.

So silk it had to be for the supporting mechanism. By July 14, a new way of supporting multiple returns of the communication line in the Great Barn had been devised: each returned length of packthread passed over a different set of horizontal silk threads. Now the defining experiment was possible. Eight returns were set up so that a line of communication of 666 feet total length was eventually attained. Much to their excitement, the leaf brass beside the receiving sphere was deflected with ease when the glass tube was rubbed. Here was evidence that a communication link of over 200 yards was possible, and still there seemed no obvious limit to the eventual length of a possible communication path.

Now they wanted to see whether the experiment would work if the communication path were in a straight line. A 650-foot length of packthread was run out of a window in the house, across the Grand Garden, and into the Great Field beyond. The communication line was supported on silk thread run between fifteen sets of poles. Gray took his glass tube into the field to produce the electric "virtue" at the remote point, while Wheler watched for the movement of the leaf brass in the house. He shouted success to Gray, as the leaf brass danced in response to the distant encouragement of the electrified glass rod. Then Wheler took charge of the tube so that Gray could see the effect of the deflected leaf brass with his own eyes. What excitement there was! But night was now approaching, and dew had started to settle. The deflection of the leaf brass became less pronounced, and by 8 P.M. had finally failed entirely. Gray pondered whether the dew was affecting the communication of the "virtue" or whether the problem was that he was now so hot through physical effort that he was failing to excite the glass tube sufficiently through rubbing.

If a single date can be assigned to the start of electrical communication, then July 14, 1729, should be it. Although there had been earlier experiments over shorter paths, and

Gray would later achieve communication over greater distances, the first Great Barn and Great Field experiments provided the breakthrough of demonstrating that there appeared to be no obvious limit to the distance over which electrical signals could be communicated. Whatever debate there might be about the impact of Gray's other work, the world of science could never be the same once the significance of the Otterden Place experiments of July 14, 1729, was established. The humble dyer had conducted one of the great experiments of science with his gentleman friend. Gray's life was finally fulfilled scientifically, against the odds of his social standing, his modest means, and Isaac Newton's suppression of his earlier work.

In the days that followed their major experiments, Gray and Wheler achieved a greater distance in a straight path by dropping a line into the garden from a turret window. By this method, communication was achieved over a path of 765 feet. They achieved an even longer communication path of 886 feet on August 1.

Although these highly significant experiments were performed during 1729, it was February 8, 1731, before Stephen Gray wrote to the Royal Society describing them. With the previous rejection of his work by the society, his reticence in communicating with them is understandable; however, he had described his results to Desaguliers on his return to London from Otterden Place. Following Gray's methods, Desaguliers had then made some simple demonstrations of communication at a meeting of the society. Gray now wanted to claim his rightful precedence:

> In the year 1729 I communicated to Dr Desaguliers and some other gentlemen a discovery I had made lately showing that the electric virtue of a glass tube may be conveyed to any other bodies, so as to give them the property of attracting and repelling light bodies, as the tube does, when excited by rubbing. This attractive virtue might be carried to bodies that were at many feet distant from the tube.

Gray's lengthy letter to the society then gave a very full account of all the communication experiments carried out during 1729 at Norton Court and Otterden Place with Godfrey's and Wheler's assistance. After decades of neglect, the Royal Society now had no hesitation publishing Gray's letter describing in full his experiments.

For a three-year period Gray, Wheler, and Godfrey enjoyed a monopoly on the study of the communication of electrical effects before others started taking forward their pioneering endeavors. Gray was ridiculed by many for suggesting that electricity was related to lightning many years before Benjamin Franklin's famous kite experiment convinced the world that this was indeed the case. Summarizing some experiments on electricity, Gray noted in 1735 that:

the electric fire which by several of these experiments seems to be of the same nature with that of thunder and lightning.

Recognition at Last

On November 25, 1731, the Royal Society had two eminent visitors: the Prince of Wales and his guest the Duke of Lorraine. The Royal Society records show that one of the demonstrations chosen for the honored guests was

electrical experiments by Mr Gray, which succeeded notwithstanding the largeness of the company. They showed the facility with which electricity passes through great lengths of conductors and are remarkable as having been the first of this nature.

A year after his electrical communication experiments were published, Gray wrote again to the society:

The approbation the former communication of my electrical experiments to the Royal Society by their most generous encouragement hath been a great inducement to me to go on with them to see what further discoveries I can make.

He successfully attempted to make the "property of electric attraction more permanent in bodies" (such as pitch, beeswax, sulfur, and resin):

The manner of preserving them in a state of attraction was by wrapping them up in anything that would keep them from external air. They were put into a large fir box there to remain until I had occasion to make use of them.

I did for 30 days continue to observe every one of these bodies and found that at the end of the said time they attracted as vigorously as at the first or second day. It will appear that some of them have not lost their attraction for more than 4 months, so that we have some reason to believe that we have discovered that there is a perpetual attractive power in all electrical bodies.

Gray had demonstrated the ability for electric charge to be stored. Simple electrical experiments were extended to a vacuum:

With a small hand air pump that was lent by a friend, I have made experiments on several bodies and find that they will attract in vacuo and that at very nearly the same distance as in pleno.

A year later, Gray submitted a third major paper to the Royal Society, called "Electrical Attraction at a Distance Without Any Contact of the Line of Communication." It demonstrated for the first time an effect now known as electrical induction—the appearance of an electrical effect in one body induced by the generation of an electrical effect in a disconnected body:

A small hoop was hung perpendicular and in a plane at right angles to the horizontal line of communication which passed through or at least very near to the centre of the hoop, then going to the end of the said line and applying the excited tube near it there was an attractive influence communicated to the hoop in all parts of it.

In a further interesting experiment, Gray used two cubes of wood with sides measuring 6 inches; one cube was solid, the other hollow. Gray linked the two wooden cubes by a length of packthread, and a glass cylinder charged by rubbing was held up to touch the center of the packthread. The attraction of leaf brass was noted at the two cubes, and Gray recorded that there was

no more attraction at the solid cube than in the hollow cube.

From this Gray concluded that the electric "virtue" must reside on the outside of an electrified object. (This important result would be acknowledged by the world of science only following experiments by Faraday a century later.)

It is worth pausing to reflect on the discoveries Stephen Gray made in the few years following 1729, when he was well into his sixties. He had demonstrated electrical communication for the first time. He had demonstrated the difference between conductors and insulators. He had revealed new classes of electrical materials, showed that electricity can be stored for long periods, and demonstrated the phenomenon of electrical induction. And he had shown that electrical effects occur in a vacuum and electric charge resides on the outside of an electrified object. No wonder Desaguliers had stated that Gray had "made greater variety of electrical experiments than all the philosophers of this and the last age." No single individual would make such a wide variety of major discoveries about electrical phenomena until Michael Faraday a century later. The long-standing effluvia theory could no longer support all Gray's results, although Gray was unable to recognize this. It would be for others to take up the challenge of producing a new theory for electricity, based on the insight provided by Gray's breakthroughs.

As previously noted, Stephen Gray was the first recipient of the Royal Society's prestigious Copley Medal in 1731. (Other recipients of the Copley Medal have included Benjamin Franklin and Albert Einstein.) The award of the medal to Gray was repeated in 1732, indicating the high regard in which he was now held by the society. He was finally nominated as a fellow on November 2, 1732. After his name had been before the society for the next ten meetings, as was the custom, he was elected a fellow on January 25, 1733, and formally admitted on March 15, 1733. He was now sixty-six years old. The form of his nomination was as follows:

Mr Stephen Gray, well known by his many curious experiments and observations, laid before this society, is proposed a candidate to be elected a Fellow thereof and recommended by us. Martin Folkes (President), Richard Graham (Fellow), Taylor White (Fellow).

Gray's sense of achievement at having achieved fellowship must have been immense. Belated and just recognition was his. All the years of toil as a dyer and trying to do what science he could after a hard day's work and with the meager resources at hand had finally paid off. His had not been the easy path of the academic scientists or the gentleman scientists, who made up the vast majority of fellows of the society. In his dealings with the ornate gathering of virtuosos of the Royal Society, Gray still showed great humility and extreme pleasure. Even allowing for the effusive language of the age, Gray's letters always showed remarkable respect. The following is a typical ending of a letter to the society:

Sir, be pleased to communicate these to the Royal Society to whom I hope they will be no less acceptable than some of my former discoveries who am, Sir, theirs and your most obedient and humble servant, Stephen Gray.

But despite Gray's new recognition, decades of neglect had meant that many of his pioneering achievements were now being taken forward by others without the proper acknowledgment of Gray's original contributions. The publication of Gray's electrical communication results in the *Philosophical Transactions* aroused the interest of Charles François de Cisternay Dufay, gardener to the king of France. He was one who did acknowledge Gray as providing the inspiration for his ideas.

Gray was overjoyed that he was now gaining respect from scientists of distinction, such as Dufay, who were taking his work forward. On January 28, 1735, Gray wrote to Cromwell Mortimer, secretary of the Royal Society:

> I see you have published Mr Dufay's letter to the Duke
> of Richmond in the *Philosophical Transactions*. It is no
> small satisfaction to me that my electrical discoveries
> have not only been confirmed by so judicious a philoso-
> pher as Mr Dufay, but yet he has made several new ones
> of his own.

Dufay's testimony to the inspiration that he had received from Stephen Gray was generous:

> He was one who worked on this subject with application
> and success, and to whom I acknowledge myself
> indebted for the discoveries I have made as well as for
> those I may possibly make hereafter, since it is from his
> writings that I took the resolution of applying myself to
> these kind of experiments.

Progress on electricity was now being made throughout Europe. It was Dufay's follow-up work and clarification and regularization of Gray's results that finally attracted scientists everywhere to the study of electricity and spelled the end of the effluvia theory. By 1733 Dufay had been able to clarify the distinction between the two types of electricity. The first, known as "vitreous," was produced by rubbing glass or rock crystal with hair or wool, and the second, known as "resinous," was produced by rubbing amber or copal with silk or paper. Benjamin Franklin later renamed the two types of electricity "positive" and "negative." Dufay also formalized the law explaining how two bodies with similar forms of electricity repelled and those with unlike forms attracted each other. (Later, the French scientist Charles-Augustin de Coulomb demonstrated that the law of electrical forces followed an inverse square relationship similar to the one Newton had established for gravity.) Meanwhile the Dutchman Pieter van Musschenbroek invented the Leyden jar to store electricity (the precursor of the modern battery). He developed a container made of a nonconductor, such as glass, coated with a conductor and filled with water. Electric charge admitted to the jar through a metal nail passing through a cork in the top could be preserved in the jar for long periods without being dissipated.

In 1747, Sir William Watson extended Gray's electrical communication demonstration in an experiment organized by the Royal Society. A conductor was run from

Westminster Bridge for a distance of 12,276 feet (almost 2.5 miles), and an electric effect was propagated along this full distance; such was the dramatic advance in the fifteen years since Gray's early communication experiments. Watson also arranged a similar demonstration for the Prince of Wales. A somewhat more amusing demonstration of the propagation of electricity was made by the Abbé Nollet before Louis XV and his court. A party of Carthusian friars was lined up along a distance of a mile, and wires linked the friars one to the other. The abbé then discharged electricity from a Leyden jar into the wire, demonstrating the passage of the electricity as all the friars leaped into the air. The electric phenomena pioneered by Gray were soon being demonstrated in the New World and aroused the interest of Benjamin Franklin, who would make his own unique contributions to the new science.

The Growth of Electrical Communication

The path leading from Stephen Gray's pioneering communication research to modern telegraphy and telephony was tortuous and uncertain. Progress was made in stuttering steps until the mid-nineteenth century, when innovations in America took over from uncertain developments in Europe. A novel proposal was made in a Scottish magazine in 1853 by an unknown author who signed himself or herself simply "C.W." C.W.'s proposal was to replicate Gray's communication experiments using twenty-six different wires—one for each letter of the alphabet. Light balls suspended at the receiving end of each of the wires would respond to the signal applied at the transmitting end. The person sending the signal could move a charged rod from wire to wire, like playing a xylophone, and the movement of the distant balls signifying letters would spell out the message at the remote location. Despite the clumsiness of the proposal, a model based on C.W.'s idea was built in Switzerland. Others realized that there was no need for a multiplicity of wires. Why not use pulsed code in the same fashion as wall tapping by medieval prisoners in isolated cells? In Italy, Alessandro Volta, inventor of the voltaic battery that provided the first reliable source of electric current, argued the case for an overhead iron-wire communication link from Lake Como to Milan.

Experimenters in Germany sent electric signals over wires and detected them by bubbles in a trough of water at the receiving end. Although this signal detection method was too slow to be other than a mere novelty, it too had been presaged by Gray's experiment showing that water would react to electricity.

In 1819 the Dane Hans Christian Ørsted detected the deflection of a compass needle adjacent to a wire carrying an electric current. Stephen Gray had noted this phenomenon, for the case of a moving charged rod, in his unpublished letter from Cambridge over a century earlier. André-Marie Ampère in France suggested that magnetic deflection could be used as a communication detector for coded signals; and in 1832 the first functional needle telegraph was constructed for the Russian czar, running between his winter and summer palaces in Saint Petersburg. In England, William Fothergill Cooke and Charles Wheatstone developed a five-needle, six-wire telegraph for sending coded messages. By July 1839, they had a working system running between Paddington Station in London and West Drayton, 18 miles away. The railway was content to work with a code of twenty letters plus ten numerals, well suited to a five-needle system for coded messages.

With the discovery that a wire coil with a soft iron core could produce a magnetic field when an electric current flowed through the coil, buzzers and other electromagnetic devices took over from the needle telegraph. At this time the center of innovation moved firmly to the United States. Samuel Morse worked on a system using a coding of signals of shorter or longer duration, and in 1835 the Morse code, a sequence of dots and dashes representing letters, numbers, and punctuation, was introduced to the world. Two years later Morse formed a partnership with Alfred Vail, and by 1843 Morse and Vail had built the first intercity overhead link, between Baltimore and Washington. Public use was inaugurated on May 24, 1844, with the somewhat frightening message "What God hath wrought." Before long telegraph lines were spanning the continents.

From 1845 to 1880 the telegraph was the standard form of communication within cities, linking police stations, fire brigades, and a network of post offices, as well as providing important links between cities. The telegraph revolutionized news communication, and as early as 1841 the Associated Press was formed so that newspapers could share their telegraph expenses. In 1865 a New York-to-Paris telegraph line route was surveyed via British Columbia, Alaska, the Bering Straits, and Siberia. The U.S. government was involved, and the plan was part of the reason for the purchase of Alaska from Russia. Construction was started eastward from Saint Petersburg and westward from British Columbia. However, the invention of insulating material suitable for submarine cables eventually halted the project, and telegraph cables were run under the Atlantic.

During the 1870s Alexander Graham Bell, Thomas Edison, and others were working on voice transmission over telegraph lines. By 1877 Bell had commercialized his inventions, and the telephone moved communications into another realm. Telegraphy, telephony, picture transmission by facsimile, and eventually data transmission linking computers would suddenly make the planet a very much smaller place and change the path of history.

The lineage of the present telecommunication revolution is clear. It can be tracked through various diverse paths, back and forth between the Old and New Worlds, to the first electrical communication experiments of Stephen Gray—one of history's forgotten heroes.

Final Days

In his final days, Gray turned his mind to a new description of the structure of the universe. It was a grandiose challenge for a scientist who had previously been content with making more subtle and cautious contributions to the advancement of science. He had given some outline of his ideas in a letter to Granville Wheler, describing an experiment whereby electricity could make small, pendulous bodies revolve around a larger central body "constantly the same way that the planets move about the sun." Wheler suggested that the revolutions were produced "by the volition of the experimenter who held the string." Yet, he noted,

> I am persuaded at the same time that he was not sensible of giving any motion to his hand himself.

It seems, sadly, that Gray's grand design was no more than the confused notions of a dying man. Nevertheless, the society's secretary, Cromwell Mortimer, was summoned to the Charterhouse to see Gray. The *Philosophical Transactions* contain "An account of some electrical experiments intended to be communicated to the Royal Society by Mr Stephen Gray FRS; taken from his mouth by Cromwell Mortimer on February 14th 1736, the day before he died."

The experiments Gray proposed to Mortimer to prove that the planets revolved around the Sun under the action of electricity were muddled nonsense, contrary to the careful approach to scientific investigation that had typified his life. Mortimer ended the paper as follows:

> He told me he had thought of these experiments only a very short time before his falling sick, that he had not yet tried them with a variety of bodies, but that from what he had already seen of them they struck him with new surprise every time he repeated them. He hoped if God would spare his life but a little longer, he should from these phenomena point out, bring his electrical experiments to a greater perfection. And he did not doubt but in a short time to be able to astonish the world with a new sort of planetarium never before thought of. From these experiments might be established a certain theory for accounting for the motions of the Grand planetarium of the Universe.

Gray died at the Charterhouse on February 15, 1736. Until the final days he did not realize that his life was nearing its end. A week before he died he had written to Granville Wheler from the Charterhouse:

I have now and then some companies of gentlemen and ladies come to me to be entertained with my electrical experiments. I am in hopes when the days are longer and the weather better I may have more than I have at present.

Scientific Epitaphs

For those wishing to honor the memory of Sir Isaac Newton, the choice of memorials is extensive. His birthplace and childhood home of Woolsthorpe Manor is open to the public as a place of homage. One can explore his boyhood graffiti scribbled on the walls, stand in the study where he performed his first light experiments with a prism, and sit in the garden where he contemplated the nature of gravity under the apple tree. (The present tree is a descendant of the original.) The nearby town of Grantham, where he went to school, has a noble statue of Newton—facing the grotesque contrast of a modern Isaac Newton shopping complex. At Cambridge University, his memory is enshrined in numerous ways, most notably a splendid statue in the antechapel of Trinity College (the Latin inscription states, "Who surpassed all men in genius"). There is also the Isaac Newton Institute, an internationally acclaimed center of mathematical excellence. His various residences in London carry commemorative plaques, and there are Newton guided tours available to his London haunts. The genius of classical science has been commemorated in painting and verse, on currency and postage stamps, and in street names the length and breadth of his country. And in Westminster Abbey, the resting place of England's greatest exponents of literature, art, science, politics, and war, stands an impressive baroque monument commemorating his life. The figure of Newton is shown leaning on books labeled "Phil Princ. Math." (the *Principia*), "Divinity," "Optica," and "Chronology." Cherubs are depicted playing with his prism, telescope, and minted coins. A celestial globe shows the path of a comet. Here, indeed, is fitting recognition of the greatest British hero of science. It confirms the image of "Saint Isaac," so carefully nurtured by his nephew-in-law, John Conduitt, and numerous subsequent biographers.

In promoting the legend of perfection for Newton, his failings were blamed on others. Thus John Flamsteed has been held liable by many Newton biographers for Newton's failure to perfect the theory of the Moon's motion. Typical of this interpretation of events is the following piece written in 1850 by J. Edleston, a fellow of Trinity College:

> In 1694 Newton renewed his attack on the lunar and planetary theories with a view to a new edition of his book (the Principia). And if Flamsteed, the Astronomer Royal, had cordially co-operated with him in the humble

capacity of an observer in the way that Newton pointed out and requested of him, (and for his almost unpardonable omission to do so I know of no better apology that can be offered than that he did not understand the real nature and, consequently, the importance of the researches in which Newton was engaged, his purely empirical and tabular views never being replaced in his mind by a clear conception of the Principle of Universal Gravitation,) the lunar theory would, if its creator did not overrate his own powers, have been completely investigated, so far as he could do it, in the first few months of 1695, and a second edition of the Principia would probably have followed the execution of the task at no long interval. But science and the world were not destined to such a good fortune. Flamsteed's infirmities of temper and bodily health conspired to thwart Newton's plans.

But Newton's headaches over the Moon were due to the complexity of the problem rather than to a shortage of data from Flamsteed. So it is a matter of mystery that history could so easily have overlooked the wealth of lunar data of good quality made available to Newton by Flamsteed. Flamsteed's monumental contributions to astronomy are no longer in doubt; the posthumous publication of *Historia Coelestis Britannica* and the *Atlas Coelestis* confirmed them for posterity.

In paying homage to John Flamsteed, there is again a large choice of sites to visit. The house in which he was born in Denby no longer stands, but a plaque on its modern replacement in Flamsteed Lane commemorates his birthplace. The Derby high school carries his name, as does a nursery school opposite Saint Bartholomew's Church at Burstow. The first Astronomer Royal lies buried with his beloved wife before the altar of the church that he had served with dedication. And in Greenwich Park, the original Wren observatory building is now called Flamsteed House. As the second millennium reached its end, the role of Greenwich in defining the reference point of longitude for the Earth was again recalled; this preeminence of Greenwich was assured by the reputation it first gained through the work of Flamsteed. Perhaps the perfect place to pay homage to the first Astronomer Royal is to stand in the beautiful Octagon Room of Flamsteed House, looking out across Greenwich Park to the river and pondering his forty-four years of dedicated labor as he mapped the heavens with a new precision. Forty-four years, thirty thousand observations, three thousand star positions: this was indeed a life dedicated to a single purpose and the unselfish giving of knowledge to humanity.

In seeking to remember Stephen Gray, the man who did more than any of his illustrious contemporaries to reveal the true nature of electricity, and to demonstrate that electrical effects could be communicated over a distance, there are no plaques, no memorials, no street names—not even a grave at which to pay homage. It is not known where Stephen Gray is buried. As a poor pensioner of the Charterhouse, he might have been given no more than an unmarked pauper's grave. His nephew, John Gray, was a successful doctor in Canterbury by the time of his uncle's death, and it could be that he returned Stephen Gray's body to be buried at All Saints Church in Best Lane where he had been christened sixty-nine years earlier. The church no longer stands. In its place there is a small garden containing a few headstones from the graves in the old churchyard; none carries the name of Gray. John Gray did make a list for the Royal Society of his uncle's meager possessions at the time of his death, a humble memorial to his achievements. Among other things, the list included

> 5 large glass tubes, 1 glass funnel with a long tube,
> 1 small glass hook in the form of an S, 6 pieces of
> coloured glass, 1 iron ball, 1 cork ball, 1 sulphur globe,
> 1 screw, 1 polished iron hoop, 2 round cakes of wax,
> 3 common hollow canes, 4 bobbins of silk, 1 iron rod
> bent with a crank in the middle.

Such were the modest tools of a humble pioneer of science.

The most moving testimony to Stephen Gray came some thirty years after his death. Anna Williams, whose father had been at the Charterhouse with Gray, wrote a poem about him. When visiting her father she helped

Gray with his electrical experiments and she claimed to have been "the first that observed and notified the emission of the electrical spark from a human body."

Anna Williams became a close friend of Dr. Samuel Johnson, the author of the first English dictionary, and he helped her with the poem about Gray that she published in 1766. The affectionate tone of the poem indicates that the humble scientist had left a lasting impression on his young helper. She was clearly unaware of the manner in which Newton had suppressed his work, but she was content to place his genius alongside Newton's, as well as Bacon's and Boyle's. In her poem, Anna Williams was able to capture the genius, humility, and values of Stephen Gray.

On the Death of Stephen Gray FRS
The author of the present doctrine of electricity

Long hast thou born the burden of the day,
Thy task is ended, venerable Gray.
No more shall art thy dextrous hand require
To break the sleep of elemental fire;
To rouse the powers that actuate Nature's frame,
The momentous shock, the electric flame,
The flame which first, weak pupil of thy lore,
I saw, condemned, alas to see no more.

Now, hoary Sage, pursue thy happy flight,
With swifter motion haste to purer light,
Where Bacon waits with Newton and with Boyle

To hail thy genius, and applaud thy toil
Where intuition breaks through time and space,
And mocks experiment's successive race;
Sees tardy Science toil at Nature's laws,
And wonders how the effect obscures the cause.

Yet not to deep research or happy guess
Is owed the life of hope, the death of peace.
Unblest the man whom philosophic rage
Shall tempt to lose the Christian in the Sage;
Not art but goodness poured the sacred ray
That cheered the parting hour of humble Gray.

Perhaps the memory of Stephen Gray, the man who "made greater variety of electrical experiments than all the philosophers of this and the last age," deserves more than merely a poem and a list of his scientific instruments. Perhaps he deserves just a small fraction of the plaques, statues, and place names used to honor Isaac Newton and John Flamsteed. But even the thought of such memorials would undoubtedly have embarrassed the humble dyer. His anonymous legacy to the world is the modern communications revolution; and the privilege of being the first to witness its early birth pangs would have provided him with far greater satisfaction than the thought of any temporal memorials.

"Thy task is ended, venerable Gray."

Bibliography

Francis Baily, *An Account of the Reverend John Flamsteed* (London 1835). There is no modern biography of Flamsteed that betters this book. Baily, working from Flamsteed's letters and other writings, was the first person to challenge the early Newton hagiographers who had denigrated the memory of the first Astronomer Royal.

David Clark and Lesley Murdin, *The Enigma of Stephen Gray: Astronomer and Scientist* (*Vistas in Astronomy*, Vol. 23, 1979). This research paper references all the letters of Stephen Gray and investigates his life and scientific advances in astronomy and electricity.

Alan Cook, *Edmond Halley: Charting the Stars* (Clarendon Press 1998). A scholarly and very readable treatise on the life and times of Edmond Halley that catches the intricacies of the relationships between Newton,

Flamsteed, and Halley, as well as Halley's many contributions to knowledge.

Margaret Espinasse, *Robert Hooke* (Heinemann 1956). Based on Hooke's diaries and correspondence, this fascinating book tracks Hooke's many contributions to science and his conflict with Newton. It gives an interesting insight into the life of seventeenth-century London.

A. Rupert Hall, *Isaac Newton: Adventure in Thought* (Cambridge University Press 1996). This is a thorough analysis of Newton's huge manuscript legacy.

Lesley Murdin, *Under Newton's Shadow: Astronomical Practices in the Late Seventeenth Century* (Adam Hilger 1985). An intriguing investigation of the lifestyles of and relationships between the astronomers of Newton's time: a mixture of professionals, such as Halley and Flamsteed, and amateurs such as Stephen Gray.

Pepys' Diaries. There is no better way of appreciating the nature of life in seventeenth-century London than to peruse the diaries of Samuel Pepys. Pepys was a well-connected naval clerk, who eventually rose to become secretary of the Admirality and the president of the Royal Society.

Dava Sobel, *Longitude* (Fourth Estate 1996). This is a thoroughly readable account of the search for a solution to the longitude problem. Clockmaker John Harrison made heroic contributions. The story mainly covers events

in the early to middle eighteenth century, so it overlaps
with the somewhat earlier events of this volume.

Richard Westfall, *Never at Rest: A Biography of Isaac
Newton* (Cambridge University Press 1980). This is the
definitive biography of Newton. Among the myriad writings on Isaac Newton, here is the balanced assessment of
the life and times of one of the greatest of Englishmen.

Michael White, *Isaac Newton: The Last Sorcerer* (Fourth
Estate 1998). To those intrigued by Isaac Newton's "dark
side" and his excursions into alchemy and religious
extremism, this book will be of interest.

Index